Nadia Nedjah, Luiza de Macedo Mourelle

Fuzzy Systems Engineering

Studies in Fuzziness and Soft Computing, Volume 181

Editor-in-chief
Prof. Janusz Kacprzyk
Systems Research Institute
Polish Academy of Sciences
ul. Newelska 6
01-447 Warsaw
Poland
E-mail: kacprzyk@ibspan.waw.pl

Further volumes of this series
can be found on our homepage:
springeronline.com

Vol. 167. Y. Jin (Ed.)
Knowledge Incorporation in Evolutionary Computation, 2005
ISBN 3-540-22902-7

Vol. 168. Yap P. Tan, Kim H. Yap,
Lipo Wang (Eds.)
Intelligent Multimedia Processing with Soft Computing, 2005
ISBN 3-540-22902-7

Vol. 169. C.R. Bector, Suresh Chandra
Fuzzy Mathematical Programming and Fuzzy Matrix Games, 2005
ISBN 3-540-23729-1

Vol. 170. Martin Pelikan
Hierarchical Bayesian Optimization Algorithm, 2005
ISBN 3-540-23774-7

Vol. 171. James J. Buckley
Simulating Fuzzy Systems, 2005
ISBN 3-540-24116-7

Vol. 172. Patricia Melin, Oscar Castillo
Hybrid Intelligent Systems for Pattern Recognition Using Soft Computing, 2005
ISBN 3-540-24121-3

Vol. 173. Bogdan Gabrys, Kauko Leiviskä,
Jens Strackeljan (Eds.)
Do Smart Adaptive Systems Exist?, 2005
ISBN 3-540-24077-2

Vol. 174. Mircea Negoita, Daniel Neagu,
Vasile Palade
Computational Intelligence: Engineering of Hybrid Systems, 2005
ISBN 3-540-23219-2

Vol. 175. Anna Maria Gil-Lafuente
Fuzzy Logic in Financial Analysis, 2005
ISBN 3-540-23213-3

Vol. 176. Udo Seiffert, Lakhmi C. Jain,
Patric Schweizer (Eds.)
Bioinformatics Using Computational Intelligence Paradigms, 2005
ISBN 3-540-22901-9

Vol. 177. Lipo Wang (Ed.)
Support Vector Machines: Theory and Applications, 2005
ISBN 3-540-24388-7

Vol. 178. Claude Ghaoui, Mitu Jain,
Vivek Bannore, Lakhmi C. Jain (Eds.)
Knowledge-Based Virtual Education, 2005
ISBN 3-540-25045-X

Vol. 179. Mircea Negoita,
Bernd Reusch (Eds.)
Real World Applications of Computational Intelligence, 2005
ISBN 3-540-25006-9

Vol. 180. Wesley Chu,
Tsau Young Lin (Eds.)
Foundations and Advances in Data Mining, 2005
ISBN 3-540-25057-3

Vol. 181. Nadia Nedjah,
Luiza de Macedo Mourelle
Fuzzy Systems Engineering, 2005
ISBN 3-540-25322-X

Nadia Nedjah
Luiza de Macedo Mourelle

Fuzzy Systems Engineering

Theory and Practice

 Springer

Nadia Nedjah
Luiza de Macedo Mourelle

Universidade do Estado do Rio de Janeiro
Faculdade de Engenharia
Rua São Francisco Xavier, 524, Sala 5022-D
Maracanã, Rio de Janeiro - RJ, 20550-900, Brasil
E-mail: nadia@eng.uerj.br
        ldmm@eng.uerj.br

Library of Congress Control Number: 2005921894

ISSN print edition: 1434-9922
ISSN electronic edition: 1860-0808
ISBN-10 3-540-25322-X Springer Berlin Heidelberg New York
ISBN-13 978-3-540-25322-8 Springer Berlin Heidelberg New York

Springer is a part of Springer Science+Business Media
springeronline.com
© Springer-Verlag Berlin Heidelberg 2005
Printed in The Netherlands

Typesetting: by the authors and TechBooks using a Springer LaTeX macro package
Cover design: E. Kirchner, Springer Heidelberg

Printed on acid-free paper   SPIN: 10984697   89/TechBooks   5 4 3 2 1 0

Preface

When asked about whether a certain aspect is or is not adequate for instance, it is politically incorrect to give a sharp answer, i.e. it is or it is not adequate. A good politician would answer that the aspect in question is adequate to some extent but it is also not adequate to another extent. So, if you want to be a successful politician, you ought to learn fuzzy logic. The mob prefers a fuzzy answer and sharp minded persons can suffer a great deal in this fuzzy old world!

Like Politics, computational system modelling is strewn with ambiguous situations, wherein the designer cannot decide, with precision, what should be the outcome of the system because of the lack of precise data on the actual situation. Lotfi Zadeh, a professor at the University of California, used the fact that human being do not have always access to precise numerical data, but they are capable of taking more or less the right the decision with respect to the situation they are in. The fuzzy theory, as it was first introduced, was an attempt to yield a new computing paradigm that allow designers to do *more with less*. At first, It was embraced in designing feedback controllers as they could be programmed to accept noisy, imprecise input. Nowadays, the fuzzy theory, including fuzzy sets, relations, functions and logic, is being exploited, in all sort of areas and disciplines.

This book is devoted to reporting innovative and significant progress in fuzzy system engineering. Theoretical as well as practical chapters are contemplated. The former present original seminal work on improving the fuzzy theory and the latter exploit it to engineer intelligent systems.

The content of this book is divided in two main parts. The chapters of the first part present novel developments of the fuzzy theory while those of the second part describe interesting applications of the fuzzy logic. In the following, we give a brief description of the main contribution of each of these chapters.

Part I: Fuzzy Theory

In Chap. 1, which is entitled *Introducing You to Fuzziness*, the authors, **Nadia Nedjah and Luiza de Macedo Mourelle**, introduce fuzzy logic and the underlying approximate reasoning. First, they present the fuzzy set theory through their operational semantics. Then, they extend the fuzzy set theory to fuzzy logic. In this purpose, we define fuzzy propositions and rules. We also demonstrate how those are used to reason approximately.

In Chap. 2, which entitled *A Qualitative Approach for Symbolic Data Manipulation Under Uncertainty*, the authors, namely **Isis Truck and Herman Akdag**, present a novel a qualitative (also symbolic and linguistic) approach for knowledge representation. They introduce a qualitative approach to manipulate uncertainty is as an alternative to classic probabilities and design a framework that supports the operational semantics of their approach.

In Chap. 3, which entitled *Adaptation of Fuzzy Inference System Using Neural Learning*, the author, namely **Ajith Abraham**, presents three different types of cooperative neuro-fuzzy models namely fuzzy associative memories, fuzzy rule extraction using self-organizing maps and systems capable of learning fuzzy set parameters. Different Mamdani and Takagi-Sugeno type integrated neuro-fuzzy systems are further introduced with a focus on some of the salient features and advantages of the different types of integrated neuro-fuzzy models that have been evolved during the last decade.

Part II: Fuzzy Systems

In Chap. 4, which is entitled *A fuzzy approach on guiding model for interception flight*, the author, namely **Silviu Ionita**, presents an original moving control model based on the fuzzy logic, applied to some navigation special issues. The author claim that the steps taken on special flight tasks prove adequacy of the fuzzy rules based model to this field.

In Chap. 5, which is entitled *Hybrid Soft and Hard Computing Based Forex Monitoring Systems*, the author, namely **Ajith Abraham**, attempts to compare the performance of hybrid soft computing and hard computing techniques to predict the average monthly forex rates one month ahead. The soft computing models considered are a neural network trained by the scaled conjugate gradient algorithm and a neuro-fuzzy model implementing a multi-output Takagi-Sugeno fuzzy inference system. The author claim that the proposed hybrid models could predict the forex rates more accurately most of the time than all the techniques when applied individually.

In Chap. 6, which is entitled *On the Stability and Sensitivity Analysis of Fuzzy Control Systems for Servo-systems*, the authors, namely **Radu-Emil Precup and Stefan Preitl**, first present the stability analysis methods dedicated to fuzzy control systems for servo-systems: the state-space approach, the use of Popov's hyperstability theory, the circle criterion and the harmonic balance method. Second, they performed the sensitivity analysis of fuzzy control systems with respect to the parametric variations of the controlled plant for a class of servo-systems based on the construction of sensitivity models. Several case studies are presented.

In Chap. 7, which is entitled *Applications of Fuzzy Logic in Mobile Robots Control*, the author, namely **Michael Botros**, presents two different approaches for the automatic design of fuzzy inference systems. The first approach is through the use of Neuro-Fuzzy architecture and a learning process to adapt the fuzzy system parameters. The second approach is through the use of genetic algorithms as an optimization tool for selecting the most suitable membership functions and rules for the fuzzy system. The author compares the two approaches and applies them to robots communication in multi-robot teams.

In Chap. 8, which is entitled *Modelling the Tennessee Eastman Chemical Process Reactor Using Fuzzy Logic*, the authors namely, **Alaa F. Sheta**, introduces the challenges associated with modeling nonlinear dynamical systems. The least square estimation method and the fuzzy method, were used to solve the modelling problem for the Tennessee Eastman chemical process reactor. The developed results show that fuzzy models can efficiently perform like the actual chemical reactor and with a high modelling capabilities.

Rio de Janeiro   *Nadia Nedjah, Ph.D.*
March 2005   *Luiza de Macedo Mourelle, Ph.D.*

Contents

Part I

Fuzzy Theory

1

Introducing You to Fuzziness

N. Nedjah[1] and L. de M. Mourelle[2]

[1] Department of Electronics Engineering and Telecommunications,
Engineering Faculty,
State University of Rio de Janeiro,
Rua São Francisco Xavier, 524, Sala 5022-D,
Maracanã, Rio de Janeiro, Brazil
nadia@eng.uerj.br, http://www.detel.eng.uerj.br
[2] Department of System Engineering and Computation,
Engineering Faculty,
State University of Rio de Janeiro,
Rua São Francisco Xavier, 524, Sala 5022-D,
Maracanã, Rio de Janeiro, Brazil
ldmm@eng.uerj.br, http://www.desc.eng.uerj.br

1.1 Introduction

Computational system modelling is full of ambiguous situations, wherein the designer cannot decide, with precision, what should be the outcome of the system. In [17], L. Zadeh introduced for the first time the concept of *fuzziness* as opposed to *crispiness* in data sets. When he invented *fuzzy sets* together with the underlying theory, Zadeh's main concern was to reduce system complexity and provide designer with a new computing paradigm that allow approximate results. Whenever there is uncertainty, *fuzzy logic* together with *approximate reasoning* apply. Fuzzy logic and approximate reasoning [18, 19] can be used in system modelling and control as well as data clustering and prediction [13], to name only few appropriate utilisations. Furthermore, they can be applied to any discipline such as finance [4], image processing [7, 16], temperature and pressure control [11, 22], robot control [9, 14], etc.

This chapter is an introduction to fuzzy logic and the underlying approximate reasoning. First, in Sect. 1.2, we present fuzzy sets through their operational semantics. Thereafter, in Sect. 1.3, we introduce fuzzy relations and explain how the composition of fuzzy relations is viewed and implemented. Then, in Sect. 1.4, the fuzzy set theory is extend to fuzzy logic. In this purpose, we define fuzzy propositions and rules. We also demonstrate how those are used to reason approximately. Subsequently, in Sect. 1.5, we describe the basic generic architecture of a fuzzy system, detailing the role each one of its component has to play. Last but not least, in Sect. 1.6, we summaries the chapter.

1.2 Fuzzy Sets

Conventionally, a set S is said to be *crisp*, if and only if given an element $x \in \mathcal{U}$, wherein $\mathcal{U}$ is called the *universe of discourse*, we either have $x \in S$ or $x \ni S$. For fuzzy sets, this no longer true. Every element $x \in \mathcal{U}$ belongs to a fuzzy set but only to some degree. A set S becomes *fuzzy*, if it is paired with a function $\mu_S : \mathcal{U} \hookrightarrow [0,1]$ called its *membership* function that, given an element $x \in \mathcal{U}$, defines its *membership degree* to S. A pair $(x, \mu_S(x))$ is called a *a singleton* and so each fuzzy set can be considered as the crisp set defined as the union of all singletons $(x, \mu_S(x))$, for all $x \in \mathcal{U}$. Note that the definition of fuzzy sets subsumes that of crisp ones as we can always write the membership function of the latter as in (1.1)

$$\mu_S \hookrightarrow \{0,1\}$$
$$\mu_S(x) = \begin{cases} 1, & \text{if } x \in S \\ 0, & \text{if } x \ni S \end{cases} \tag{1.1}$$

There many example of crisp vs. fuzzy sets in the literature. Perhaps the best example to illustrate the need of fuzziness is when one wants to define the concept of *cold* and *cool* water. Crisp definitions of these three concepts may be those in (1.2).

$$\mu_{cold}(t) = \begin{cases} 1, & \text{if } -10°C \leq t \leq 0°C \\ 0, & \text{otherwise} \end{cases}$$

$$\mu_{cool}(t) = \begin{cases} 1, & \text{if } 0°C \leq t \leq 10°C \\ 0, & \text{otherwise} \end{cases} \tag{1.2}$$

As we can see in the graphical representation of the crisp definitions of (1.2) in Fig. 1.1, the concepts of cold and cool water are mutually exclusive.

Certainly, one would expect that water at temperature $-10°C$ is much colder than that at $1°C$. Water at temperature $-10°C$ is rather frozen than cold. Analogically, water at temperature $10°C$ is much cooler than that at $0°C$. Water at temperature $10°C$ is rather warm that cool. On the other hand, it is not accurate to define a transition from cold to cool by the application of a single degree Celsius. In the real world, one would expect a gradual and smooth drift between these two states. The stated considerations are not taken account of in the crisp definitions in (1.2).

It is common to use graphics to describe the membership function of fuzzy sets. For the sake of simplicity, *triangular*, *trapezoidal*, *s-shaped* and *Gauss*

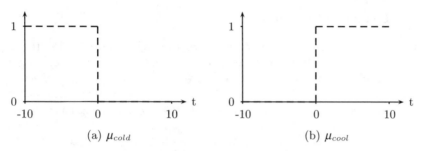

(a) μ_{cold} (b) μ_{cool}

Fig. 1.1. Crisp representation of the concepts *cold* and *cool* water

graphics are preferred. The triangular fuzzy representation of the concepts of
cold and cool water can be defined as in (1.3) and depicted graphically as in
Fig. 1.2.

$$\mu_{cold}(t) = \begin{cases} \frac{2}{15}t + \frac{5}{3}, & \text{if } -12.5°C < t \leq -5°C \\ -\frac{2}{15}t + \frac{1}{3}, & \text{if } -5°C < t \leq 2.5°C \\ 0, & \text{otherwise} \end{cases}$$

$$\mu_{cool}(t) = \begin{cases} \frac{2}{15}t + \frac{1}{3}, & \text{if } -2.5°C < t \leq 5°C \\ -\frac{2}{15}t + \frac{5}{3}, & \text{if } 5°C < t \leq 12.5°C \\ 0, & \text{otherwise} \end{cases}$$

$$(1.3)$$

Considering the graph of Fig. 1.2, we can see that water with temperature
$0°C$ is $\frac{1}{3}$ cold and as much, i.e. $\frac{1}{3}$ cool. To allow a smooth drift from one
concept o another, e.g. form cold to cool, a superposition of 10 to 50% is
usually used. In the example of Fig. 1.2, we used a superposition is of about
33%. Note that the points at which the membership function of a fuzzy set
is 0.5 are called *crossover points* of the set. In Fig. 1.2, the highlighted points
are the crossover points of the fuzzy sets *cold* and *cool* respectively.

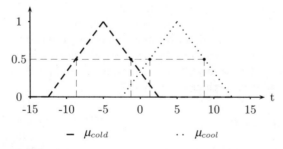

$-\ \mu_{cold}$ $\cdot\cdot\ \mu_{cool}$

Fig. 1.2. Fuzzy representation of the concepts *cold* and *cool* water

1.2.1 Linguistic Variables and Terms

Two main aspects, essential to fuzzy system modelling, are [17, 19]: the concepts to be quantified such as *temperature, age* and *height* and the characteristics of the identified concepts such as *cold* and *cool* for temperature, *young* and *old* for age and *short* and *tall* for height. The concepts quantified within a fuzzy system are called *linguistic variables* while the corresponding characteristics are called *linguistic terms* [6].

In fuzzy systems, each linguistic variable of interest $V \in \mathcal{R} \subseteq \mathcal{U}$, where $\mathcal{R}$ is called the variable referential set which is a subset of the universe of discourse $\mathcal{U}$ is quantified using a set of linguistic terms $T_V^1, T_V^2, \ldots, T_V^n$. Each linguistic term T_V^i is actually a fuzzy set having its own membership function $\mu_{T_V^i}$.

For instance, consider the linguistic variable *temperature* of tank water. Let *cold, cool, warm* and *hot* be the characteristics of the water we are interested in. Possible membership functions of these characteristics are shown in the graph of Fig. 1.3. The referential set of the linguistic variable is $[-15, 35]$.

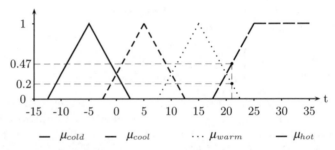

Fig. 1.3. Fuzzy representation of linguistic variable *temperature* using *cold, cool, warm* and *hot* linguistic terms

1.2.2 Operators

In order to operate on linguistic variables, one needs some operators on fuzzy sets. These fuzzy set operators are often an extension of their counterparts for crisp or conventional sets. In the following, we define the fuzzy version of set operations as well as their usefulness.

Fuzzy Set Intersection

There are many ways to perform the intersection of two fuzzy sets. Intersection operators that satisfy reasonable axioms for a truth-functional definition of intersection are called triangular norms (or *T-norms*). Let A and B be two fuzzy sets defined by their respective membership functions μ_A and μ_B

respectively. Operator $\sqcap$ such that $\mu_{A \cap B}(x) = \mu_A(x) \sqcap \mu_B(x)$ is said to be a T-norm if and only if for all elements in $[0,1]$, it satisfies the following T-norm axioms:

- $\sqcap: [0,1] \times [0,1] \rightarrow [0,1]$;
- Commutativity: $\forall x, \mu_A(x) \sqcap \mu_B(x) = \mu_B(x) \sqcap \mu_A(x)$;
- Associativity: $\forall C, x, (\mu_A(x) \sqcap \mu_B(x)) \sqcap \mu_C(x) = \mu_A(x) \sqcap (\mu_B(x) \sqcap \mu_C(x))$;
- Monotonicity: $\forall \alpha', \beta', x, \mu_A(x) \geq \alpha', \mu_B(x) \geq \beta' \iff$
  $\mu_A(x) \sqcup \mu_B(x) \geq \alpha' \sqcup \mu_B(x), \mu_A(x) \sqcup \mu_B(x) \geq \mu_A(x) \sqcup \beta'$;
- Boundary conditions: $\mu_A(x) \sqcap 1 = \mu_A(x), \mu_A(x) \sqcap 0 \doteq 0$.

Intuitively, the elements of the fuzzy set obtained as the intersection of two other fuzzy sets representing two distinct characteristics are expected to be those elements that present both characteristics (to some extent) at the same time. The most common intersection operators for fuzzy sets are *Zadeh intersection* or *standard*, *product intersection* or *algebraic*, *Lukasiewicz intersection* or *limited* and *robust*. These operators are defined in (1.4), (1.5), (1.6) and (1.7). As an example, we plotted (see Fig. 1.4) the intersection of the membership functions μ_{cold} and μ_{cool} of Fig. 1.2 using the standard and algebraic operators. Note that both the limited and robust intersection operators produce a membership function $\mu_{cold \cap cool}(x) = 0$.

$$(\mu_A \sqcap \mu_B)(x) = \min(\mu_A(x), \mu_B(x)) \tag{1.4}$$

$$(\mu_A \sqcap \mu_B)(x) = \mu_A(x) \cdot \mu_B(x) \tag{1.5}$$

$$(\mu_A \sqcap \mu_B)(x) = \max(\mu_A(x) + \mu_B(x) - 1, 0) \tag{1.6}$$

$$(\mu_A \sqcap \mu_B)(x) = \begin{cases} \mu_A(x), & \text{if } \mu_B(x) = 1 \\ \mu_B(x), & \text{if } \mu_A(x) = 1 \\ 0, & \text{otherwise} \end{cases} \tag{1.7}$$

The Cartesian product $A \times B$ of two fuzzy sets A and B is associated with the membership function $\mu_{A \times B}(a, b) = \min(\mu_A(a), \mu_B(b))$ with $a \in A$

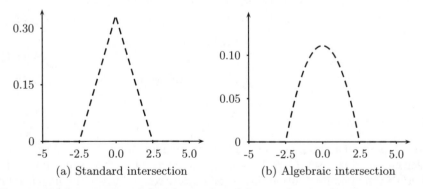

(a) Standard intersection   (b) Algebraic intersection

Fig. 1.4. Representation of the intersection of the membership functions μ_{cold} and μ_{cool} of Fig. 1.2 using the standard and algebraic operators

and $b \in B$. Note that $A \times B$ coincides with $A \sqcap B$ when using the standard intersection operator.

Fuzzy Set Union

As for intersection, there are many ways to perform the union of two fuzzy sets. Union operators that satisfy reasonable axioms for a truth-functional definition of union are called triangular conorms (or *T-conorms*). Let A and B be two fuzzy sets defined by their respective membership functions μ_A and μ_B respectively. Operator $\sqcup$ such that $\mu_{A \cup B}(x) = \mu_A(x) \sqcup \mu_B(x)$ is said to be a T-conorm if and only if for all elements in [0,1], it satisfies the following T-conorm axioms. Observe that the first four axioms are the same for T-norm ones. The boundary conditions, however, are complementary.

- $\sqcup: [0,1] \times [0,1] \rightarrow [0,1]$;
- Commutativity: $\forall x, \mu_A(x) \sqcup \mu_B(x) = \mu_B(x) \sqcup \mu_A(x)$;
- Associativity: $\forall C, x, (\mu_A(x) \sqcup \mu_B(x)) \sqcup \mu_C(x) = \mu_A(x) \sqcup (\mu_B(x) \sqcup \mu_C(x))$;
- Monotonicity: $\forall \alpha', \beta', x, \mu_A(x) \geq \alpha', \mu_B(x) \geq \beta' \iff$
  $\mu_A(x) \sqcup \mu_B(x) \geq \alpha' \sqcup \mu_B(x), \mu_A(x) \sqcup \mu_B(x) \geq \mu_A(x) \sqcup \beta'$;
- Boundary conditions: $\mu_A(x) \sqcup 1 = 1, \mu_A(x) \sqcup 0 = \mu_A(x)$.

Intuitively, the elements of the fuzzy set obtained as the union of two other fuzzy sets representing two distinct characteristics are expected to be those elements that present either characteristic (to some extent) as well as those that present both of them (to some extent). The most common union operators for fuzzy sets are *Zadeh union* or *standard*, *product union* or *algebraic*, *Lukasiewicz union* or *limited* and *robust*. These operators are defined in (1.8), (1.9), (1.10) and (1.11). As an example, we plotted (see Fig. 1.5) the union of the membership functions μ_{cold} and μ_{cool} of Fig. 1.2 using the given union operators.

$$(\mu_A \sqcap \mu_B)(x) = \max(\mu_A(x), \mu_B(x)) \tag{1.8}$$

$$(\mu_A \sqcap \mu_B)(x) = \mu_A(x) + \mu_B(x) - \mu_A(x) \cdot \mu_B(x) \tag{1.9}$$

$$(\mu_A \sqcap \mu_B)(x) = \min(1, \mu_A(x) + \mu_B(x)) \tag{1.10}$$

$$(\mu_A \sqcap \mu_B)(x) = \begin{cases} \mu_A(x), & \text{if } \mu_B(x) = 0 \\ \mu_B(x), & \text{if } \mu_A(x) = 0 \\ 1, & \text{otherwise} \end{cases} \tag{1.11}$$

Fuzzy Set Complement

As intersection and union operators, there are many ways to perform the complement of a fuzzy set. Complement operators need to satisfy some axioms for a truth-functional definition. Let A be a fuzzy set defined by its membership function μ_A. Unary operator, say η, is said to be a complement operator if and only if for all elements in [0,1], it satisfies the following axioms.

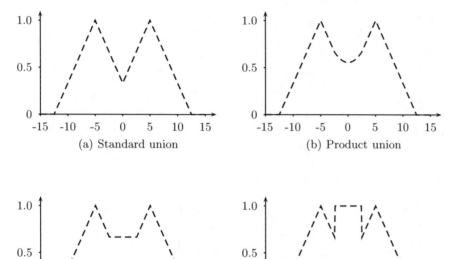

Fig. 1.5. Representation of the union of the membership functions μ_{cold} and μ_{cool} of Fig. 1.2 using most common operators

- η: $[0,1] \rightarrow [0,1]$;
- Boundary: $\eta(0) = 1$ and $\eta(1) = 0$;
- Monotonicity: $\forall x, y, \ \mu_A(x) < \mu_A(y) \ Longleftrightarrow \ \eta(\mu_A(x)) > \eta(\mu_A(y))$
- Involution: $\forall x, \eta(\eta(\mu_A(x))) = \mu_A(x)$

Intuitively, the elements of the fuzzy set obtained as the complement of another fuzzy set representing a given characteristic are expected to be those elements in the referential set that do not present that characteristic (to some extent). The most common complement operators for fuzzy sets are the *standard complement*, *Sugeno complement* and *Yager complement*. These operators are defined in (1.12), (1.13) and (1.14). As an example, we plotted (see Fig. 1.6) the complement of the union of the membership functions μ_{cold} and μ_{cool} of Fig. 1.2 using the given complement operators.

$$\eta(\mu_A(x)) = 1 - \mu_A(x) \tag{1.12}$$

$$\eta_{sigma}(\mu_A(x)) = \frac{1 - \mu_A(x)}{1 + \sigma \times \mu_A(x)} \tag{1.13}$$

$$\eta_v(\mu_A(x)) = (1 - \mu_A(x)^v)^{\frac{1}{v}} \tag{1.14}$$

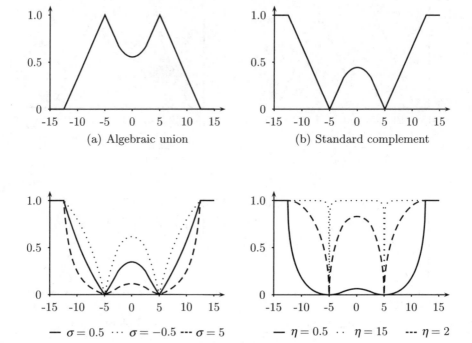

Fig. 1.6. Representation of the complement of the algebraic union membership functions μ_{cold} and μ_{cool} of Fig. 1.2 using most common operators

Modifiers

This kind of operators is used in an attempt to model situation where one needs to concentrate or dilate a given fuzzy term using hedges such as "very" and "fairly". This is a very handy propriety in Fuzzy system modelling [10]. Given a characteristic A with membership function μ_A, the hedge "very" intensifies the characteristic associating as a membership function $\mu_{very\ A}(x) = \mu_A^2(x)$ while the hedge "fairly" dilates the characteristic using as a membership function $\mu_{faily\ A}(x) = \sqrt{\mu_A(x)}$. The linguistic terms *very cold* and *fairly cold* with respect to the concept *cold* of Fig. 1.2 are represented in Fig. 1.7.

Analogically, the definition of other more specific modifiers, such as "extremely", "slightly" and even "exactly", can be easily derived form that of hedges "very" and "slightly". For instance, modifier "extremely" can be defined as $\mu_{extremely\ A}(x) = \mu_A^3(x)$ and modifier "slightly" as $\mu_{slightly\ A}(x) = \mu_A^{1/3}(x)$. Furthermore, modifier "exactly" can be implemented as the highest intensification of a given characteristic and so can be modelled as $\lim_{p\to\infty} \mu_{exactely\ A}(x) = \mu_A^p(x)$.

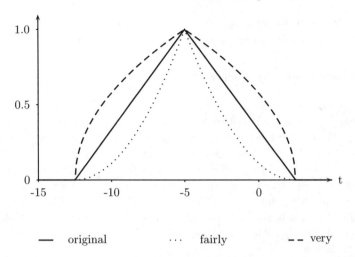

— original   ⋯ fairly   -- very

Fig. 1.7. Representation of the modifications caused by the hedge operators

1.3 Fuzzy Relation

A *crisp relation* $R^c : A \hookrightarrow B$ establish whether element $a \in A$ is related
to element $b \in B$ with respect to relation R^c. This determines either pair
$(a, b) \in R^c$ or $(a, b) \ni R^c$. So a crisp relation can be viewed as a crisp set
of all pairs (a, b) in the Cartesian product $A \times B$ such that a is related to b
with respect to relation R^c. As one can guess, a *fuzzy relation* R^f establish
the degree to which an element $a \in A$ is related to element $b \in B$ with
respect to relation R^f. So a fuzzy relation can be viewed as a crisp set of
all pairs $((a, b), \mu^{R_f}(a, b))$ where (a, b) is any pair of $A \times B$. Fuzzy relation
$R^f : A \hookrightarrow B$ can also be viewed as a $|A| \times |B|$ matrix with same name
wherein $R^f_{ij} = \mu_R(i, j)$.

Composition of relations is computed using the so-called *min-max* rule
[20, 21]. Let $\bullet$ be the composition operator of fuzzy relation. Assume that
$R : A \hookrightarrow B$ and $S : B \hookrightarrow C$ be two fuzzy relations. The membership function
μ_T of fuzzy relation $T : A \hookrightarrow C$ such that $T = R \bullet S$ is computed as in (1.15)
for all $a \in A$ and $c \in C$.

$$\mu_T(a, c) = \max_{b \in B}(\min(\mu_R(a, b), \mu_S(b, c))) \qquad (1.15)$$

1.4 Fuzzy Logic

Fuzzy logic [1, 17] attempts to quantify the degree of truth of propositions
and consequences. Unlike *crsip logic* where either a proposition is true or
false, in *fuzzy logic*, a proposition may be true (1), half true (0.5) or false
(0). This gives way to *multi-valued* logic and *approximate* reasoning instead

of conventional *two-valued* logic and *precise* reasoning. One can also consider a fine-gain multi-valued logic considering 3/4 and 1/4 truth.

1.4.1 Fuzzy Logic Connectives

As for crisp propositions, *fuzzy propositions* may be simple are composed, i.e. connecting two or more propositions. A simple fuzzy proposition is usually of the form x is T, wherein x is linguistic variable and T is a linguistic term. For instance, *temperature* is *hot*, *age* is *young* or *height* is *tall*.

For composition of propositions, one uses *fuzzy connectives* as opposed to conventional crisp, which are *negation* or $\neg$, *conjunction* or $\wedge$, *disjunction* or $\vee$, *implication*, or $\Rightarrow$ and *equivalence* or $\Leftrightarrow$. Let $P(x)$ and $Q(y)$ be two fuzzy propositions which have the truth degree $\mu_P(x)$ and $\mu_Q(y)$ respectively, with $x \in \mathcal{R}_P$ and $y \in \mathcal{R}_Q$. The degree of truth yield by these fuzzy connectives is defined as follows:

- negation: $\mu_{\neg P}(x) = 1 - \mu_{P(x)}$;
- conjunction: $\mu_{P \wedge Q}(x,y) = \min(\mu_{P(x)}, \mu_Q(y))$;
- disjunction: $\mu_{P \vee Q}(x,y) = \max(\mu_P(x), \mu_Q(y))$;
- implication: $\mu_{P \Rightarrow Q}^G(x,y) = \max((\mu_P(x) \leq \mu_Q(y)), \mu_Q(y))$;
- equivalence: $\mu_{P \Leftrightarrow Q}^G(x,y) = \max((\mu_P(x) == \mu_Q(y)), \min(\mu_P(x), \mu_Q(y)))$.

Note that the fuzzy definitions of the implication and equivalence connectives are based on Gögel's. However, as this connective is very important for fuzzy inference and define the outcome of the on lying approximate reasoning, there has been many other implication definitions [2]. A non-exhaustive list of possible interpretations of a fuzzy implication are given below. Some of these operators contradict some of the theorems in crips logic [8].

- Mamdani's: $\mu_{P \Rightarrow Q}^M(x,y) = \min(\mu_P(x), \mu_Q(y))$;
- Zadeh's: $\mu_{P \Rightarrow Q}^Z(x,y) = \max(\min(\mu_P(x), \mu_Q(y)), (1 - \mu_P(x)))$;
- Yager's: $\mu_{P \Rightarrow Q}^Y(x,y) = \mu_Q(y)^{\mu_P(x)}$;
- Larsen's: $\mu_{P \Rightarrow Q}^L(x,y) = \mu_P(x).\mu_Q(y)$;
- Gaines's: $\mu_{P \Rightarrow Q}^G(x,y) = \max((\mu_P(x) \leq \mu_Q(y)), \mu_Q(y)/\mu_P(x))$;
- Lukasiewicz's: $\mu_{P \Rightarrow Q}^K(x,y) = \min(1, 1 - \mu_P(x) + \mu_P(x).\mu_Q(y)$.

1.4.2 Fuzzy Rules and Inference

A fuzzy *rule* is simply an implication which antecedent and consequent fuzzy propositions. Implications are usually expressed as *if-then* statements rather than using the implication logic symbol ($\Rightarrow$). So proposition $P(x) \Leftrightarrow Q(y)$ is commonly written as if $P(x)$ then $Q(y)$.

Traditionally, *inference* is known as the process that draws conclusions form a set of facts using a collection of (crisp) rules. A rule fires when its antecedent proposition is true. This simply the application of the *modus*

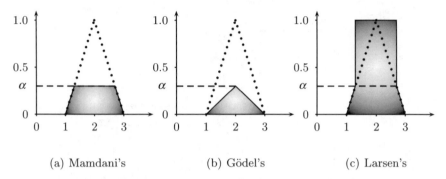

(a) Mamdani's   (b) Gödel's   (c) Larsen's

Fig. 1.8. Representation of the result of some implication operators

ponens theorem that establish that for propositions P and Q, we have $(P \wedge (P \Rightarrow Q)) \Rightarrow Q$. When a conclusion is reached, it is then sharply true. In contrast with the described traditional inference process, fuzzy inference is based on fuzzy rules and applies the same *modus ponens* principle. However, as the rules and the facts are fuzzy, when a conclusion is reached from the application of rule if $P(x)$ then $Q(y)$, it only true to a certain degree of truth, which is computed $\mu_{P \Leftrightarrow Q}(x, y)$ [5]. The graphical representation of some the implication operators is depicted in Fig. 1.8.

1.5 Fuzzy Controllers

Fuzzy control, which directly uses fuzzy rules is the most important and common application of the fuzzy theory [15]. Using a procedure originated by E. Mamdani [12], three steps are taken to design a fuzzy controlled machine:

1. fuzzification or encoding: This step in the fuzzy controller is responsible of encoding the crisp measured values of the system parameter into a fuzzy term using the respective membership functions;
2. inference: This step consists of identifying the subset of fuzzy rules that can be fired, i.e. those with antecedent propositions with truth degree not zero, and draw the adequate fuzzy conclusions;
3. defuzzification or decoding: This is the reverse process of fuzzification. It is responsible of decoding a fuzzy variable and compute its crisp value.

The generic architecture of a fuzzy controller is given in Fig. 1.9. Its main components consist of a *knowledge repository*, the *encoder* or *fuzzifier*, the *decoder* or *defuzzifier* and the *inference engine*. The knowledge base stores two kind of data: the fuzzy rules which are required by the inference engine to reach the expected results and knowledge about the fuzzy terms together with their respective membership functions as well as information about the universe of discourse of each fuzzy variable manipulated within the controller. The

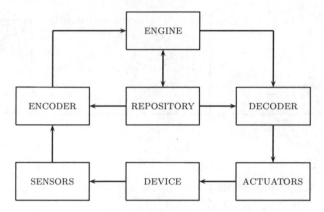

Fig. 1.9. Generic architecture of fuzzy controllers

encoder implements the transformation from crip to fuzzy and the decoder the transformation from fuzzy to crisp. Of course, the inference engine is the main component of the controller architecture. It implements the approximate reasoning process.

1.5.1 Operation

In order to explain thoroughly how a fuzzy controller operates, we borrow the famous example of a the inverted pendulum from [3]. The inverted pendulum is described in Fig. 1.10. The problem consists of controlling the movement of a pole on a mobile platform. It can only balance to the right or left.

The linguistic variables are the *angle* between the platform and the pendulum ($[-30°, 30°]$), the angular *velocity* of this angle ($[-15,15]$) and the *speed* of the platform ($[-3, 3]$). The first two variables are the controller input while

Fig. 1.10. Inverted Pendulum

Table 1.1. Fuzzy variable and corresponding linguistic terms

Angle – A	Velocity – V	Speed – S
n-large	n-high	n-fast
n-small	n-low	n-slow
insignificant	null	still
p-small	p-low	p-slow
p-large	p-high	p-fast

the third one is the expected controller output. The fuzzy terms for each of the identified variables are given in Table 1.1. The membership functions of the controller linguistic fuzzy terms *angle*, *velocity* and *speed* are given in Fig. 1.11, Fig. 1.12 and Fig. 1.13 respectively.

The collection of the fuzzy rules, which is used to control the pendulum, is given in a tabular form in Table 1.2. The controller rules are generally designed by an expert to achieve optimal control. For instance, the third entry in the third row reads as the rule stated in (1.16) while the second entry of the fourth row reads as the rule stated in (1.17).

$$\text{if } angle \text{ is } insignificant \land velocity \text{ is } null \text{ then } speed \text{ is } still \qquad (1.16)$$

$$\text{if } angle \text{ is } p\text{-}small \land velocity \text{ is } n\text{-}low \text{ then } speed \text{ is } still \qquad (1.17)$$

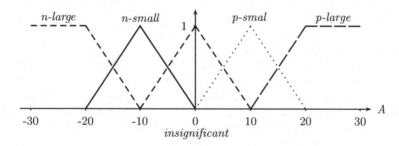

Fig. 1.11. Fuzzy representation of membership function μ_{angle}

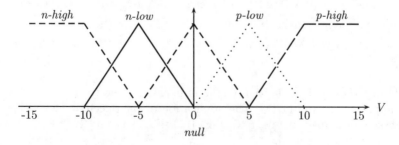

Fig. 1.12. Fuzzy representation of membership function $\mu_{velocity}$

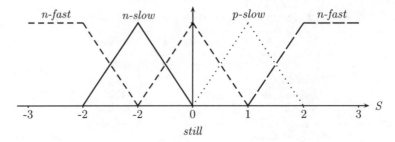

Fig. 1.13. Fuzzy representation of membership function μ_{speed}

Table 1.2. Fuzzy rules of the controller

Rules	n-large	n-small	Insignificant	p-small	p-large
n-high	–	–	n-fast	–	–
n-low	–	–	n-slow	still	–
null	n-fast	n-slow	still	p-slow	p-fast
p-low	–	still	p-slow	–	–
p-high	–	p-fast	–	–	–

Now, we show how to apply the rules of Table 1.2 for specific measures of the linguistic variables *angle* and *velocity*. Assume that the value read for variable *angle* is 6° and that for variable *velocity* is −1. The values of the membership functions for variable *angle* are $\mu_{insignificant}(16) = 0.4$ and $\mu_{p\text{-}small}(16) = 0.6$ while for variable *velocity* $\mu_{null}(-2) = 0.2$ and $\mu_{p\text{-}low}(-2) = 0.8$. These points are marked on the graphs of Fig. 1.14.

The rules that applies are those that have their degree of truth different from zero. So we conclude that all rules with antecedent involving the linguistic terms *insignificant* and/or *p-small* and *null* and/or *n-low* should be fired. From Table 1.2, we can identify that the rules that should be used are those whose consequent is framed. The four rules are also listed below in (1.18).

if (*angle* is *insignificant*) ∧ (*velocity* is *null*)   then *speed* is *still*
if (*angle* is *insignificant*) ∧ (*velocity* is *n-low*)   then *speed* is *n-slow*
if (*angle* is *p-small*) ∧ (*velocity* is *n-low*)   then *speed* is *still*
if (*angle* is *p-small*) ∧ (*velocity* is *null*)   then *speed* is *p-slow*

$$(1.18)$$

Using Mamdani's definition of the implication operator (see Fig. 1.8, we can apply he rules in (1.18). The application of the selected rules yields the fuzzy sets described in Fig. 1.15, Fig. 1.16, Fig. 1.17 and Fig. 1.18. Always the minimum cut is used.

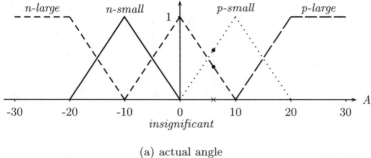

(a) actual angle

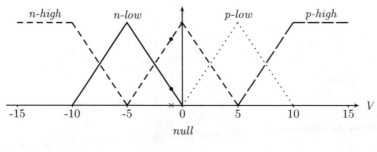

(b) actual velocity

Fig. 1.14. Membership reading for the actual values of variable *angle* and *velocity*

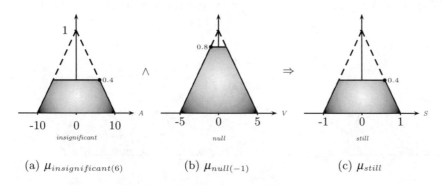

(a) $\mu_{insignificant(6)}$ (b) $\mu_{null(-1)}$ (c) μ_{still}

Fig. 1.15. Representation of the result of application of the first rule in (1.18)

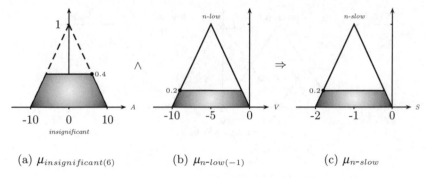

(a) $\mu_{insignificant(6)}$ (b) $\mu_{n\text{-}low(-1)}$ (c) $\mu_{n\text{-}slow}$

Fig. 1.16. Representation of the result of application of the second rule in (1.18)

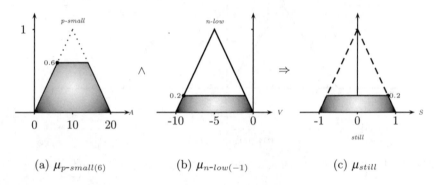

(a) $\mu_{p\text{-}small(6)}$ (b) $\mu_{n\text{-}low(-1)}$ (c) μ_{still}

Fig. 1.17. Representation of the result of application of the third rule in (1.18)

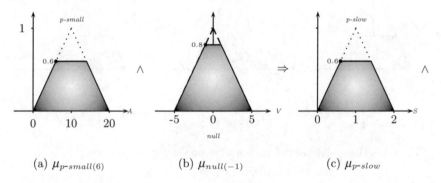

(a) $\mu_{p\text{-}small(6)}$ (b) $\mu_{null(-1)}$ (c) $\mu_{p\text{-}slow}$

Fig. 1.18. Representation of the result of application of the fourth rule in (1.18)

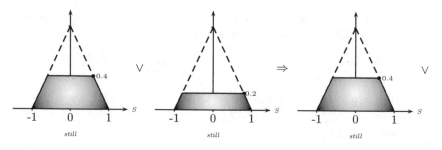

(a) Result of first rule   (b) Result of third rule   (c) Final result

Fig. 1.19. Combining the results of application of the first and third rule in (1.18)

As the first and third rule in the selected rules yield the same consequent, i.e. *still*, we can combine them using an or-operator and therefore, the maximum cut is retained. This is depicted in Fig. 1.19.

The defuzzification of the obtained result is a very sensitive task. It depends on the nature of the process being controlled. There are several techniques for obtaining a crisp value from a fuzzy set. However, the commonly used techniques are of two kind and are given below. Each of these techniques is illustrated in Fig. 1.20.

- the *composite moments* or *centroid*, which computes the crisp measure u as the abscissa of the centre of gravity of the obtained fuzzy set. This computation for defuzzification is described in (1.19);

$$u = \frac{\sum_i \mu(x_i) x_i}{\sum_i \mu(x_i)} \tag{1.19}$$

- the *composite maximum*, which is based on the values of the fuzzy set with the highest degree of truth. The composite maximum technique may use:
  - average maximum, which is the average of all the values that have the highest membership degree in the obtained fuzzy set; The computation

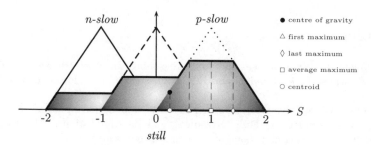

Fig. 1.20. Composing the results of all fired rules

for defuzzification is described in (1.20), wherein N represents the number of points with maximum membership degree in the fuzzy sets;

$$u = \frac{\sum_i \{x|\ \mu(x) = \max(x_i)\}}{N} \qquad (1.20)$$

– first maximum, which is the smallest value that has the highest membership degree in the obtained fuzzy set; The computation for defuzzification is described in (1.21)

$$u = \{x|\ \mu(x) = \max(x_i) \wedge \forall x_i, x < x_i\} \qquad (1.21)$$

– last maximum, which is the biggest value that has the highest degree of truth in the yield fuzzy set; The computation for defuzzification is described in (1.22)

$$u = \{x|\ \mu(x) = \max(x_i) \wedge \forall x_i, x > x_i\} \qquad (1.22)$$

1.6 Summary

This chapter is an introduction to fuzziness including fuzzy set theory, fuzzy relations, fuzzy logic and the underlying approximate reasoning. First, we presented fuzzy sets through their operational semantics. We defined usual operations on fuzzy sets, which include fuzzy set intersection, union and complement. All these operations are illustrated through graphical examples. Thereafter, we introduced fuzzy relations and explain how the composition of fuzzy relations is viewed and implemented using the min-max principle. Then, we showed how the fuzzy set theory can be extended to fuzzy logic. In this purpose, we define fuzzy propositions, rules, inference and approximate reasoning as opposed to precise reasoning. Subsequently, we described the basic generic architecture of a fuzzy controllers, detailing the role each one of its component has to play. We concluded using a thorough example to explain how fuzzy logic is used in process control.

References

1. Baldwin, J.F., Fuzzy logic and fuzzy reasoning, In Fuzzy Reasoning and Its Applications, E.H. Mamdani and B.R. Gaines (Eds.), London Academic Press, 1981.
2. Bandler, W. and Kohout, L.J., Semantics of implication operators and fuzzy relational products, In Fuzzy Reasoning and Its Applications, E.H. Mamdani and B.R. Gaines (Eds.), London Academic Press, 1981.
3. Bernstein, D., Control tutorial for MatLab: inverted pendulum, University of Michigan, USA, http://www.engin.umich.edu/group/ctm/examples/pend/invpen.html, December 2004.

4. Diao, Y., Hellerstein, J. and Parekh, S., Using fuzzy control to maximize profits in service level management, IBM Systems Journal, vol. 41, no. 3, pp. 403–420, 2002.
5. Eschbach, M. and Cunnyngham, J., The logic of fuzzy Bayesian influence, International Fuzzy Systems Association Symposium of Fuzzy information Processing in Artificial Intelligence and Operational Research, Cambridge, England, 1984.
6. Esragh, F. and Mamdani, E.H., A general approach to linguistic approximation, in Fuzzy Reasoning and Its Applications, E.H. Mamdani and B.R. Gaines (Eds.), London Academic Press, 1981.
7. Franke, K., Köppen, M. and Nickolay, B., Fuzzy image processing by using Dubois and Prade fuzzy norm, In Proceedings of 15th International Conference on Pattern Recognition, Barcelona, Spain, pp. 518–521, 2000.
8. Fox, J., Towards a reconciliation of fuzzy logic and standard logic, International Journal of Man-Machine Studies, vol. 15, pp. 213–220, 1981.
9. Ghidary, S., Hattori, M., Tadokoro, S. and Takamori, T., Multi-modal human robot interaction for map generation. Proceedings of IEEE International Conference on Intelligent Robots and Systems, pp. 2246–2251, 2001.
10. Haack, S., Do we need fuzzy logic?, International Journal of Man-Machine Studies, vol. 11, pp. 437–445, 1979.
11. Magdalena, L. and Velasco, J.R., Fuzzy Rule-Based Controllers that Learn by Evolving their Knowledge Base, In Herrera F. and Verdegay J.L. (Eds.), Genetic Algorithms and Soft Computing, Physica-Verlag, pp. 172–201, 1996.
12. Mamdani, E.H. and Assilian, S., An experiment in linguistic synthesis of fuzzy controllers, International Journal of Man-Machine Studies, no. 7, pp. 1–13, 1975.
13. Radecki, T., An evaluation of the fuzzy set theory approach to information retrieval, in R. Trappl, N.V. Findler, and W. Horn (Eds.), Progress in Cybernetics and System Research, vol. 11, Proceedings of a Symposium Organized by the Austrian Society for Cybernetic Studies, Hemisphere Publishing Company, NY. 1982.
14. Shim, D.H., Koo, T.J., Hoffmann, F. and Sastry, S.S., A comprehensive study of control design for an autonomous helicopter, Proceedings of 37th. Conference on Decision and Control, Tampa, FL, pp. 3653–3658, 1998.
15. Umbers, I.G. and King, P.J., An analysis of human decision-making in cement kiln control and the implications for automation, International Journal of Man-Machine Studies, vol. 12, pp. 11–23, 1980.
16. Wachs, J., Stern, H. and Edan, Y., Parameter Search For An Image Processing Fuzzy Cmeans, Proceedings of IEEE International Conference on Image Processing, Barcelona, Spain, vol. 3, pp. 341–344, 2003.
17. Zadeh, L.A., Fuzzy Sets , Journal of Information and Control, vol. 8, pp. 338-353, 1965.
18. Zadeh, L.A., Fuzzy algorithms, Information and Control, vol. 12, pp. 94–102, 1968.
19. Zadeh, L.A., Making computers think like people, IEEE Spectrum, no. 8, pp. 26–32, 1984.
20. Zadeh, L.A., Fuzzy Logic, IEEE Computer Journal, vol. 1, no. 83, p. 18, 1988.
21. Zimmermann, H.J. and Zysno, P., Latent connectives in human decision making, Fuzzy Sets and Systems, vol. 4, pp. 37–51, 1980.
22. Zhang, J. and Knoll, A., Designing Fuzzy Controllers by Rapid Learning, Fuzzy Sets and Systems, 101, pp. 287–301, 1999.

A Qualitative Approach
for Symbolic Data Manipulation
Under Uncertainty

I. Truck[1] and H. Akdag[2]

[1] Labo IA, University Paris 8,
   2, rue de la Liberté, 93526 SAINT-DENIS Cedex 2, France
   truck@ai.univ-paris8.fr, www.ai.univ-paris8.fr/~truck
[2] LIP6, University P. & M. Curie,
   8, rue du Capitaine Scott, 75015 PARIS, France
   LERI, University of Reims, rue des Crayères, 51687 REIMS Cedex 2, France
   herman.akdag@lip6.fr, www-apa.lip6.fr/LOFTI/Equipe/Herman/

In decision making problems, the experts knowledge has different natures: numerical, interval-valued, symbolic, linguistic,... In other terms, the information is heterogeneous. In this chapter, a qualitative (also symbolic and linguistic) approach for knowledge representation is presented. It is the continuation of studies in a given many-valued logic. First of all, a qualitative approach to manipulate uncertainty is presented as an alternative to classic probabilities. After having defined the linguistic counterpart of the symbolic framework, *Generalized Symbolic Modifiers* are defined to modify symbolic data usually expressed with linguistic terms. In the context, qualitative data are represented by degrees on a totally ordered scale. As fuzzy modifiers that act on kernels and supports of membership functions, several kinds of symbolic modifiers acting on scales and degrees are proposed. The other object of this work is data combination, especially when data do not have the same importance. The combination operator, called the *Symbolic Weighted Median*, deals with weights and proposes a representative answer depending on an initial piece of data. One interesting point is that this new median is constructed on the *Generalized Symbolic Modifiers*.

2.1 Introduction

In many expertise domains, human reasoning deals usually with linguistic concepts rather than with precise, interval-valued or fuzzy numbers. In this case, the nature of the information is qualitative and the use of structured symbols is common and suitable.

The aim of this chapter is to present different tools manipulating these symbols as well as human reasoning handles natural linguistic statements. Moreover such a reasoning is often performed under uncertainty and imprecision. That's why in order to simulate or to automatize expert reasoning, it is necessary to study the representation and treatment of discrete and linguistic data [1–5]. This chapter constitutes both a summary of previous works and a continuation of other different works on symbolic data [1, 6–8], and more precisely on vague knowledge representation (imprecise and/or uncertain): it is part of studies in many-valued logic framework.

The many-valued logic, which is a generalization of classical boolean logic, introduces truth degrees which are intermediate between *true* and *false* and allows us to represent the partial truth notion. This logic is specified by the use of $\mathcal{L}_M = \{\tau_0, \ldots, \tau_i, \ldots, \tau_{M-1}\}$[1] a totally ordered finite set[2] of truth-degrees ($\tau_i \leq \tau_j \Leftrightarrow i \leq j$) between τ_0 (false) and τ_{M-1} (true), given the operators $\vee$ (max), $\wedge$ (min) and $\neg$ (negation or symbolic complementation, with $\neg\tau_j = \tau_{M-j-1}$) and the following Lukasiewicz implication $\rightarrow_L$:

$$\tau_i \rightarrow_L \tau_j = \begin{cases} \tau_{M-1} & \text{if } i \leq j \\ \tau_{M-1-(i-j)} & \text{if } i > j \end{cases}$$

It is to notice that this implication can be written more synthetically:

$$\tau_i \rightarrow_L \tau_j = \min(\tau_{M-1}, \tau_{M-1-(i-j)})$$

These degrees can be seen as membership degrees: x partially belongs to a multiset[3] A with a degree τ_i if and only if $x \in_{\tau_i} A$. This many-valued logic can be easily applied to knowledge representation. More precisely, it deals with linguistic statements of the following form: x is $v_\alpha A$ where x is a variable, v_α a scalar adverb (such as "very", "more or less", etc.) and A a graduable linguistic predicate (such as "tall", "hot", "young"...). The predicate A is satisfiable to a certain degree expressed through the scalar adverb v_α. De Glas proposed the following interpretation [9]:

$$x \text{ is } v_\alpha A \Leftrightarrow \text{``} x \text{ is } A\text{'' is } \tau_\alpha\text{--true}$$

When knowledge is uncertain and not quantified, qualitative degrees constitute a good way to represent it because they can be associated with Zadeh's linguistic variables [10] that model well approximative reasoning.

In this framework, several qualitative approaches for uncertainty representation have been presented in the literature. Among these, [11–13] present

---

[1] with M a positive integer not null.
[2] $\mathcal{L}_M$ can be also considered as a *scale*.
[3] A multiset is a generalization of a set: in a multiset, elements may *partially* belong to it (to a certain degree).

approaches where the investigation is motivated by finding a model which simulates some activities of a cognitive agent, such as management of uncertain
statements of natural language which are defined in a finite totally ordered set
of symbolic values. The approach, as probabilities do with numerical values,
consists in representing and exploiting the uncertainty by *qualitative* degrees.
These degrees can be interpreted as "symbolic" and "comparative" degrees of
uncertainty. In order to manipulate these symbolic values, four elementary operators are constructed: multiplication, addition, subtraction and division [14].
These operators presented in the next section are based on Lukasiewicz implication and on the symbolic complementation to express the negation. In this
theoretical framework a rigorous model for evaluating uncertainties in a qualitative manner is defined. The proposed axioms and the derived properties
detailed further can be used in Bayesian networks as well as in knowledge-
based systems, in the lack of reliability quantitative estimation. However, in
these works, a fixed number of degrees is used (e.g. the set of seven degrees,
$\mathcal{L}_7$, in [4]). Thus, in a reasoning process, if the result of a cognitive operator
is between two consecutive degrees, then an approximation is needed. More
generally, one can not express a piece of data with more precision (or less)
than the granularity of the scale, at this stage. One goal of the present work
is to propose tools to solve this point. Indeed qualitative modifiers that permit the refinement of a symbolic variable, are defined: this way, data can be
represented with the needed precision.

In this last case, degrees are totally ordered as usual and they constitute a *scale*. Tools to compare and combine these degrees are introduced: the
generalized symbolic modifiers. They act on degrees as well as on scales themselves and they permit to modify data at will, i.e. to reach the needed level
of precision.

Another problem may also be raised when dealing with totally ordered
sets in practical studies such as opinion polls: the weights associated to the
symbolic degrees. In [9,13] ... the underlying condition is that all the degrees
have the same importance, i.e. the same weight. But in many situations, one
can have recourse to weights to express the relative importance of an expert
opinion, or to express the confidence that one has in data, etc. In these situations how to give a representative opinion or value? For example, how to
find the neutral value, in other words, the center of gravity? In the current
study a tool is proposed to answer this question: the *symbolic weighted median* that gives the emergent element of a set of ordered weighted degrees. The
introduced operator is constructed using the generalized symbolic modifiers.
Indeed, the *symbolic weighted median* gives as a representative element, an
element from the initial set, but slightly or strongly *modified*.

The chapter is organized as follows: in the first part (Sect. 2) a qualitative
approach for the management of uncertainty is detailed and the associated
axiomatic is reminded. The third section presents a new proposal for the
symbolic modifiers introduced in [15] and generalized in [16]: these symbolic
operators act on scale degrees and scales themselves and they allow us to

modify data at will and to associate linguistic terms with modifications. Then in Sect. 4 a new tool is proposed for the expression of the emergent value of a weighted ordered set. This operator is based on the use of the *generalized symbolic modifiers* and takes into account the weights: it is the *symbolic weighted median*. Finally, Sect. 5 concludes this study.

2.2 Symbolic Data Representation

Following the work of Akdag, De Glas & Pacholczyk [1] on the representation of the uncertainty via the many-valued logic, the first step of the approach is the proposition of an axiomatic for symbolic probability theory [1]. Khoukhi [7] and Akdag & Khoukhi [2] went further in the field of representation uncertainty and imprecision via the many-valued logic. They have proposed several solutions for the definition of each symbolic operator. Using theoretical results of [7] and [2], Seban [17] gave a convincing example of practical application in pattern recognition. Similarly, Pacholczyk & Pacholczyk [18] have proposed a new version of this theory that seems not totally satisfactory.

The proposed qualitative uncertainty theory is situated between the classical probability theory and possibility theory. The problematic here is quite close to the work of López de Mántaras in [19]. Well-chosen formulas permit to easily translate the four basic operations (represented through synthetic tables) respecting required properties. Adequate qualities of the tables are strengthened by results (theorems) of the proposed Axiomatic. All the tables are obtained thanks to logical formulas that use the implication of Lukasiewicz and the operations addition, subtraction, min and max. Thus, one can tell that the qualitative operations can be computed with the four symbolic operators, operating on integers, and this constitutes a link between the qualitative and the numerical model.

The aforementioned works [1, 2, 7, 18, 19] have in common the following points:

1. They use non-numerical degrees of uncertainty. Indeed, it is more frequent that the human expert expresses his knowledge by using qualitative approximation of numerical values.
2. They are based on the association "probability degrees/logic": in their work, laws of probabilities are obtained thanks to logical operators which are associated to truth tables.
3. They have both developed an axiomatic theory, which allows to obtain results either from axioms, or from theorems.
4. In addition, they permit the use of uncertainty (and imprecision) expressed in the qualitative form.

The considered qualitative degrees of uncertainty belong to the graduated scale $\mathcal{L}_M$. For sake of simplicity, all the examples in this section are in $\mathcal{L}_7$. To each degree of the scale, a linguistic value can be associated:

$\tau_0 \Leftrightarrow$ IMPOSSIBLE
$\tau_1 \Leftrightarrow$ NEARLY-IMPOSSIBLE
$\tau_2 \Leftrightarrow$ SLIGHTLY-POSSIBLE
$\tau_3 \Leftrightarrow$ POSSIBLE
$\tau_4 \Leftrightarrow$ RATHER-POSSIBLE
$\tau_5 \Leftrightarrow$ VERY-POSSIBLE
$\tau_6 \Leftrightarrow$ CERTAIN

The implication of Lukasiewicz is also used as well as the symbolic complementation $\neg$. In order to be able to rewrite (or to translate in symbolic) the different axioms and theorems of the classical probability theory[4], and after having introduced the concept of total order in the scale of degrees, it is easy to see as in [12] that it is necessary to define a symbolic addition (or a symbolic t-conorm, to generalize the addition), a symbolic multiplication (or a symbolic t-norm in order to be able to translate the disconditioning and the independence) and a symbolic division (to translate the conditioning).

A supplementary constraint is introduced: *"it is necessary that if C is the result of the division of A by B then B multiplied by C gives A"* (relationship between the qualitative multiplication and the qualitative division). This constraint which seems very intuitive, has been proposed for the first time in [20]. Another originality lies in the definition of a qualitative subtraction instead of the use (sometimes artificially) of the complementation operator. Thus, a symbolic difference is defined to translate the subtraction.

Beyond the translation of different axioms and theorems of classical probability theory, this approach permits, as in [18] to focus on the uncertainty of a logical implication and on the generalized modus ponens.

Authors [18] have defined a C–Independence between events and [20] define the quasi-independence of events this way: "Two events are quasi-independent if it is possible to perform the probability of their intersection by a symbolic multiplication". In the following, the model (operators + axiomatic + principal theorems) described in [8, 13, 14, 20] is detailed.

2.2.1 Formulas for Uncertainty Qualitative Theory

A qualitative multiplication of two degrees τ_α and τ_β is defined by the function MUL from $\mathcal{L}_M \times \mathcal{L}_M$ to $\mathcal{L}_M$ that verifies the following properties:

---

[4] Especially the translation of:
$P(\emptyset) = 0,\ P(\Omega) = 1,\ P(\neg A) = 1 - P(A),$
$P(A \cup B) = P(A) + P(B) - P(A \cap B),$
$P(A \cap B) = P(B/A) \times P(A)$ and $P(B/A) = P(A \cap B)/P(A).$

M1. $\text{MUL}(\tau_\alpha, \tau_0) = \tau_0$ Absorbent element
M2. $\text{MUL}(\tau_\alpha, \tau_{M-1}) = \tau_\alpha$ Neutral element
M3. $\text{MUL}(\tau_\alpha, \tau_\beta) = \text{MUL}(\tau_\beta, \tau_\alpha)$ Commutativity
M4. $\text{MUL}(\tau_\alpha, \text{MUL}(\tau_\beta, \tau_\gamma)) =$ Associativity
        $\text{MUL}(\text{MUL}(\tau_\alpha, \tau_\beta), \tau_\gamma)$
M5. $\text{MUL}(\tau_\alpha, \tau_\beta) \leq \text{MUL}(\tau_\gamma, \tau_\delta)$ Monotony
        if $\tau_\alpha \leq \tau_\gamma$ and $\tau_\beta \leq \tau_\delta$
M6. $\text{MUL}(\tau_\alpha, \neg\tau_\alpha) = \tau_0$ Complementarity

M2 to M5 are properties of a t-norm [21] to which are added the absorbent element and the complementarity properties.

Proposition 1. $\text{MUL}_{\mathcal{L}}(\tau_\alpha, \tau_\beta) = \neg(\tau_\alpha \rightarrow_L \neg\tau_\beta) = \tau_\gamma$
            therefore $\tau_\gamma = \max(\tau_{\alpha+\beta-(M-1)}, \tau_0)$

It corresponds in $\mathcal{L}_7$ to Table 2.1.

Table 2.1. Multiplication in $\mathcal{L}_7$

τ_α τ_β	τ_0	τ_1	τ_2	τ_3	τ_4	τ_5	τ_6
τ_0	τ_0	τ_0	τ_0	τ_0	τ_0	τ_0	τ_0
τ_1	τ_0	τ_0	τ_0	τ_0	τ_0	τ_0	τ_1
τ_2	τ_0	τ_0	τ_0	τ_0	τ_0	τ_1	τ_2
τ_3	τ_0	τ_0	τ_0	τ_0	τ_1	τ_2	τ_3
τ_4	τ_0	τ_0	τ_0	τ_1	τ_2	τ_3	τ_4
τ_5	τ_0	τ_0	τ_1	τ_2	τ_3	τ_4	τ_5
τ_6	τ_0	τ_1	τ_2	τ_3	τ_4	τ_5	τ_6

Similarly, a qualitative addition of two degrees τ_α and τ_β is a function ADD from $\mathcal{L}_M \times \mathcal{L}_M$ to $\mathcal{L}_M$ that verifies the following properties:

A1. $\text{ADD}(\tau_\alpha, \tau_{M-1}) = \tau_{M-1}$ Absorbent element
A2. $\text{ADD}(\tau_\alpha, \tau_0) = \tau_\alpha$ Neutral element
A3. $\text{ADD}(\tau_\alpha, \tau_\beta) = \text{ADD}(\tau_\beta, \tau_\alpha)$ Commutativity
A4. $\text{ADD}(\tau_\alpha, \text{ADD}(\tau_\beta, \tau_\gamma)) = \text{ADD}(\text{ADD}(\tau_\alpha, \tau_\beta), \tau_\gamma)$ Associativity
A5. $\text{ADD}(\tau_\alpha, \tau_\beta) \leq \text{ADD}(\tau_\gamma, \tau_\delta)$ if $\tau_\alpha \leq \tau_\gamma$ and   Monotony
        $\tau_\beta \leq \tau_\delta$
A6. $\text{ADD}(\tau_\alpha, \neg\tau_\alpha) = \tau_{M-1}$ Complementarity

A2 to A5 are properties of a t-conorm [21] to which are added the absorbent element and the complementarity properties.

Proposition 2. $\text{ADD}_{\mathcal{L}}(\tau_\alpha, \tau_\beta) = \neg(\tau_\alpha \rightarrow_L \neg\tau_\beta) = \tau_\delta$ *therefore* $\tau_\delta = \min(\tau_{\alpha+\beta}, \tau_{M-1})$

Table 2.2. Addition in $\mathcal{L}_7$

τ_α \ τ_β	τ_0	τ_1	τ_2	τ_3	τ_4	τ_5	τ_6
τ_0	τ_0	τ_1	τ_2	τ_3	τ_4	τ_5	τ_6
τ_1	τ_1	τ_2	τ_3	τ_4	τ_5	τ_6	τ_6
τ_2	τ_2	τ_3	τ_4	τ_5	τ_6	τ_6	τ_6
τ_3	τ_3	τ_4	τ_5	τ_6	τ_6	τ_6	τ_6
τ_4	τ_4	τ_5	τ_6	τ_6	τ_6	τ_6	τ_6
τ_5	τ_5	τ_6	τ_6	τ_6	τ_6	τ_6	τ_6
τ_6	τ_6	τ_6	τ_6	τ_6	τ_6	τ_6	τ_6

It corresponds in $\mathcal{L}_7$ to Table 2.2.

The qualitative subtraction of two degrees τ_α and τ_β such that $\tau_\beta \leq \tau_\alpha$ is defined by the function SOUS from $\mathcal{L}_M \times \mathcal{L}_M$ to $\mathcal{L}_M$ that verifies the following properties:

S1. $\text{SOUS}(\tau_\alpha, \tau_0) = \tau_0$ Absorbent element

S2. $\text{SOUS}(\tau_\alpha, \tau_\gamma) \leq \text{SOUS}(\tau_\beta, \tau_\gamma)$ if $\tau_\alpha \leq \tau_\beta$ Increasing relatively to the first argument

S3. $\text{SOUS}(\tau_\alpha, \tau_\beta) \leq \text{SOUS}(\tau_\alpha, \tau_\gamma)$ if $\tau_\beta \leq \tau_\gamma$ Decreasing relatively to the second argument

S4. $\text{SOUS}(\tau_\alpha, \tau_\alpha) = \tau_0$

Proposition 3. $\text{SOUS}_\mathcal{L}(\tau_\alpha, \tau_\beta) = \neg(\tau_\alpha \rightarrow_L \neg\tau_\beta) = \tau_S$ *therefore* $\tau_S = \max(\tau_{\alpha-\beta}, \tau_0)$

It corresponds in $\mathcal{L}_7$ to Table 2.3.

SOUS allows us to define an important axiom linking probability of the union to the probability of the intersection (see U6). SOUS corresponds in fact to the bounded difference of Zadeh, defined in the fuzzy logic framework [21].

Table 2.3. Subtraction in $\mathcal{L}_7$

τ_α \ τ_β	τ_0	τ_1	τ_2	τ_3	τ_4	τ_5	τ_6
τ_0	τ_0						
τ_1	τ_1	τ_0					
τ_2	τ_2	τ_1	τ_0				
τ_3	τ_3	τ_2	τ_1	τ_0			
τ_4	τ_4	τ_3	τ_2	τ_1	τ_0		
τ_5	τ_5	τ_4	τ_3	τ_2	τ_1	τ_0	
τ_6	τ_6	τ_5	τ_4	τ_3	τ_2	τ_1	τ_0

The qualitative division of two degrees τ_α and τ_β and such that $\tau_\alpha \leq \tau_\beta$ with $\tau_\beta \neq \tau_0$ is defined by the function DIV from $\mathcal{L}_M \times \mathcal{L}_M$ to $\mathcal{L}_M$ that verifies the following properties:

D1. $\text{DIV}(\tau_0, \tau_\alpha) = \tau_0$ Absorbent element

D2. $\text{DIV}(\tau_\alpha, \tau_{M-1}) = \tau_\alpha$ Neutral element

D3. $\text{DIV}(\tau_\alpha, \tau_\beta) \leq \text{DIV}(\tau_\gamma, \tau_\delta)$ if $\tau_\alpha \leq \tau_\gamma$ Increasing relatively to the first argument

D4. $\text{DIV}(\tau_\alpha, \tau_\beta) \geq \text{DIV}(\tau_\alpha, \tau_\gamma)$ if $\tau_\beta \leq \tau_\gamma$ Decreasing relatively to the second argument

D5. If $\tau_\alpha \neq \tau_0$ then[5] $\text{DIV}(\tau_\alpha, \tau_\alpha) = \tau_{M-1}$ Boundary conditions

Proposition 4. $\text{DIV}_\mathcal{L}(\tau_0, \tau_\beta) = \text{MUL}_\mathcal{L}(\tau_0, \tau_\beta)$ *if* $\tau_\beta \neq \tau_0$
$\text{DIV}_\mathcal{L}(\tau_\alpha, \tau_\beta) = \text{ADD}_\mathcal{L}(\tau_\alpha, \neg\tau_\beta)$ *if* $\tau_\alpha \leq \tau_\beta$ *and* $\tau_\beta \neq \tau_0$.

It corresponds in $\mathcal{L}_7$ to Table 2.4.

Table 2.4. Division in $\mathcal{L}_7$

τ_β \ τ_α	τ_0	τ_1	τ_2	τ_3	τ_4	τ_5	τ_6
τ_0							
τ_1	τ_0	τ_6					
τ_2	τ_0	τ_5	τ_6				
τ_3	τ_0	τ_4	τ_5	τ_6			
τ_4	τ_0	τ_3	τ_4	τ_5	τ_6		
τ_5	τ_0	τ_2	τ_3	τ_4	τ_5	τ_6	
τ_6	τ_0	τ_1	τ_2	τ_3	τ_4	τ_5	τ_6

Compared to the table proposed in [18], all ambiguities are removed (the result of a division is a unique value) and the link between the division and the multiplication actually exists, in other words:

If $\tau_x = \text{DIV}_\mathcal{L}(\tau_\alpha, \tau_\beta)$ Then $\tau_\alpha = \text{MUL}_\mathcal{L}(\tau_\beta, \tau_x)$

2.2.2 Axiomatic for Qualitative Uncertainty Theory

Considering the qualitative uncertainty concept introduced by aforementioned authors, a new axiomatic is given in [20] by using the predicate **Inc** of a many-valued logic. Thus, the proposition "A is τ_α–certain" is translated into $Inc(A) = \tau_\alpha$. The first axioms[6] are those proposed in [1]:

---

[5] where $\text{DIV}(\tau_0, \tau_0)$ is undetermined.

[6] They allow us to translate: $P(\emptyset) = 0, P(\Omega) = 1, P(\neg A) = 1 - P(A)$.

U1. If $\models_{\tau_{M-1}} A$ then $Inc(A) = \tau_{M-1}$ i.e. if A is a tautology then A is certain

U2. If $\models_{\tau_0} A$ then $Inc(A) = \tau_0$ i.e. if A is a false then A is impossible

U3. If $\models_{\tau_\alpha} A$ then $Inc(\neg A) = \neg \tau_\alpha$ i.e. if A is τ_α–certain then $\neg A$ is $\neg \tau_\alpha$–certain

U4. If $A \equiv B$ then $Inc(A) = Inc(B)$ i.e. if A is equivalent to B then $Inc(A) = Inc(B)$

This permits to have trivial theorems:

T1. If $Inc(A) = \tau_{M-1}$ then $Inc(\neg A) = \tau_0$
T2. If $Inc(A) = \tau_0$ then $Inc(\neg A) = \tau_{M-1}$

The following axioms are more original[7]:

Definition 1. *Let A and B be two quasi-independent events. If* $\mathrm{Inc}(A) = \tau_\alpha$ *and* $\mathrm{Inc}(B) = \tau_\beta$ *then* $\mathrm{Inc}(A \cap B) = \tau_\gamma$ *where* $\tau_\gamma = MUL_{\mathcal{L}}(\tau_\alpha, \tau_\beta) = \max(\tau_{\alpha+\beta-(M-1)}, \tau_0)$

U5. If $Inc(A) = \tau_\alpha$ and $Inc(B) = \tau_\beta$ and $Inc(A \cap B) = \tau_0$
then $Inc(A \cup B) = \tau_\delta$ with $\tau_\delta = ADD_{\mathcal{L}}(\tau_\alpha, \tau_\beta)$
U6. If $Inc(A) = \tau_\alpha$ and $Inc(B) = \tau_\beta$ and $B \subset A$
then $Inc(A \cap \neg B) = \tau_\sigma$ with $\tau_\sigma = SOUS_{\mathcal{L}}(\tau_\alpha, \tau_\beta)$
U7. If $Inc(A) = \tau_\alpha$ and $Inc(B) = \tau_\beta$ and $Inc(A \cap B) = \tau_\gamma$
then $Inc(A \cup B) = \tau_\theta$ with $\tau_\theta = \theta_{\mathcal{L}}(\tau_\alpha, \tau_\beta) = ADD_{\mathcal{L}}(SOUS_{\mathcal{L}}(\tau_\alpha, \tau_\gamma), \tau_\beta)$
U8. If $Inc(A \cap B) = \tau_\gamma$ and $Inc(A) = \tau_\alpha$
then $Inc(B/A) = \tau_\zeta$ with $\tau_\zeta = DIV_{\mathcal{L}}(\tau_\gamma, \tau_\alpha)$

Notice the originality of axiom (U7) where it is necessary to perform first the subtraction then the addition. One should not perform: $SOUS_{\mathcal{L}}(ADD_{\mathcal{L}}(\tau_\alpha, \tau_\beta), \tau_\gamma)$.

Remark: $ADD_{\mathcal{L}}(SOUS_{\mathcal{L}}(\tau_\alpha, \tau_\gamma), \tau_\beta)$ is equivalent to $ADD_{\mathcal{L}}(SOUS_{\mathcal{L}}(\tau_\beta, \tau_\gamma), \tau_\alpha)$.

2.2.3 Formal Consequences of This Axiomatic

The goal of this axiomatic is to show that the chosen symbolic operators permit to demonstrate[8] properties[9] that can be compared to the classical results of probability theory[10]. These results can be used in inferential process (in Knowledge Based Systems) as well as in Bayesian networks.

---

[7] Especially, they allow us to translate: $P(A \cup B) = P(A) + P(B)$ if $P(A \cap B) = \emptyset$, $P(A \cup B) = P(A) + P(B) - P(A \cap B)$, $P(B/A) = P(A \cap B)/P(A)$.
[8] The first 17 demonstrations are elementary [13], that is why they are omitted.
[9] in bold characters.
[10] in normal characters.

P1. **If $Inc(\mathbf{A}) = \tau_\alpha$ then $Inc(\mathbf{A} \cap \neg\mathbf{A}) = \tau_0$ and $Inc(\mathbf{A} \cup \neg\mathbf{A}) = \tau_{M-1}$**
$\mathrm{Prob}(A \cap \neg A) = 0$ and $\mathrm{Prob}(A \cup \neg A) = 1$

P2. **If $Inc(\mathbf{A} \cap \mathbf{B}) = \tau_\gamma$ and $Inc(\mathbf{A}) = \tau_\alpha$ and $Inc(\mathbf{B}) = \tau_\beta$**
then $Inc(\neg(\mathbf{A} \cap \mathbf{B})) = \neg\tau_\gamma$ and $Inc(\neg\mathbf{A} \cup \neg\mathbf{B})) = \neg\tau_\gamma$ One of the De
Morgan formula.

P3. **If $Inc(\mathbf{A} \cup \mathbf{B}) = \tau_\delta$ and $Inc(\mathbf{A}) = \tau_\alpha$ and $Inc(\mathbf{B}) = \tau_\beta$**
then $Inc(\neg(\mathbf{A} \cup \mathbf{B})) = \neg\tau_\delta$ and $Inc(\neg\mathbf{A} \cap \neg\mathbf{B})) = \neg\tau_\delta$ One of the De
Morgan formula.

P4. **If $Inc(\mathbf{A}) = \tau_\alpha$ and $Inc(\mathbf{B}) = \tau_\beta$ and $\mathbf{B} \subset \mathbf{A}$ then $Inc(\mathbf{B}) \leq Inc(\mathbf{A})$**
If $B \subset A$ then $\mathrm{Prob}(B) \leq \mathrm{Prob}(A)$

P5. **If $\mathbf{A} \subset \mathbf{B}$ and $\mathbf{B} \subset \mathbf{A}$ then $\mathbf{A} \equiv \mathbf{B}$ with $Inc(\mathbf{A}) = Inc(\mathbf{B})$**
If $A \subset B$ and $B \subset A$ then $\mathrm{Prob}(A) = \mathrm{Prob}(B)$

P6. **If $Inc(\mathbf{A}) = \tau_\alpha$ and $Inc(\mathbf{B}) = \tau_\beta$ with $\alpha + \beta \leq M - 1$**
and $Inc(\mathbf{A} \cap \mathbf{B}) = \tau_0$
then $Inc(\mathbf{A} \cup \mathbf{B}) = \tau_\delta$ with $\delta = \alpha + \beta$
If $\mathrm{Prob}(A \cap B) = 0$ then $\mathrm{Prob}(A \cup B) = \mathrm{Prob}(A) + \mathrm{Prob}(B)$

P7. **If $Inc(\mathbf{A}) = \tau_\alpha$ and $Inc(\mathbf{A} \cap \mathbf{B}) = \tau_0$ and $Inc(\mathbf{A} \cup \mathbf{B}) = \tau_\delta$ with $\alpha \leq \delta$**
then $Inc(\mathbf{B}) = \tau_\beta$ with $\beta = \delta - \alpha$
If $\mathrm{Prob}(A \cap B) = 0$ then $\mathrm{Prob}(B) = \mathrm{Prob}(A \cup B) - \mathrm{Prob}(A)$

P8. **If $Inc(\mathbf{A}) = \tau_\alpha$ and $Inc(\mathbf{B}) = \tau_\beta$ with $\alpha + \beta > M - 1$**
then $Inc(\mathbf{A} \cap \mathbf{B}) = \tau_\gamma$ with $\gamma = \alpha + \beta - (M - 1)$ and $Inc(\mathbf{A} \cup \mathbf{B}) = \tau_\theta$ with $\theta = M - 1$
If $\mathrm{Prob}(A \cap B) \neq 0$ then $\mathrm{Prob}(A \cup B) = \mathrm{Prob}(A) + \mathrm{Prob}(B) - \mathrm{Prob}(A \cap B)$

P9. **If $Inc(\mathbf{A}) = \tau_0$ and $Inc(\mathbf{B}) = \tau_\alpha$ then $Inc(\mathbf{A} \cap \mathbf{B}) = \tau_0$**
with $Inc(\mathbf{A} \cup \mathbf{B}) = \tau_\alpha$
If $\mathrm{Prob}(A) = 0$ then $\mathrm{Prob}(A \cap B) = 0$ and $\mathrm{Prob}(A \cup B) = \mathrm{Prob}(B)$

P10. **If $Inc(\mathbf{A}_k) = \tau_{\alpha k}$ with $k = 1..n$ and $n \geq 2$**
then $Inc(\mathbf{A}_1 \cap \mathbf{A}_2 \cap \ldots \cap \mathbf{A}_n) = \tau_{\gamma n}$
with $\gamma n = \begin{cases} \mathbf{MUL}_{\mathcal{L}}(\tau_{\alpha(n-1)}, \tau_{\alpha n}) & \text{if } n = 2 \\ \mathbf{MUL}_{\mathcal{L}}(\tau_{\gamma(n-1)}, \tau_{\alpha n}) & \text{if } n > 2 \end{cases}$
In the classical probability theory, this property corresponds to the generalization of a conjunction of n observations.

P11. **If $Inc(\mathbf{A}_k) = \tau_{\alpha k}$ with $k = 1..n$ and $n \geq 2$ and $Inc(\mathbf{A}_k \cap \mathbf{A}_l) = \tau_0 \forall k \neq l$**
then $Inc(\mathbf{A}_1 \cup \mathbf{A}_2 \cup \ldots \cup \mathbf{A}_n) = \tau_{\delta n}$
with $\delta n = \begin{cases} \mathbf{ADD}_{\mathcal{L}}(\tau_{\alpha(n-1)}, \tau_{\alpha n}) & \text{if } n = 2 \\ \mathbf{ADD}_{\mathcal{L}}(\tau_{\delta(n-1)}, \tau_{\alpha n}) & \text{if } n > 2 \end{cases}$
In the classical probability theory, this property corresponds to the generalization of a disjunction of n observations (δ–additivity of the probability).

P12. **If $\mathbf{A}$ and $\mathbf{B}$ are two quasi–independent events and $Inc(\mathbf{A}) = \tau_\alpha$ and $Inc(\mathbf{B}) = \tau_\beta$ and $Inc(\mathbf{A} \cap \mathbf{B}) = \tau_\gamma$**
then $Inc(\mathbf{B}/\mathbf{A}) = Inc(\mathbf{B}) = \tau_\beta$

If A and B are two independent events then $\text{Prob}(B/A) = \text{Prob}(B)$ or $\text{Prob}(A/B) = \text{Prob}(A)$

P13. **If $Inc(\mathbf{B/A}) = \tau_\zeta$ and $Inc(\mathbf{A}) = \tau_\alpha$**
then $Inc(\mathbf{A} \cap \mathbf{B}) = \tau_\gamma = \mathbf{MUL}_{\mathcal{L}}(Inc(\mathbf{B/A}), Inc(\mathbf{A}))$
$\text{Prob}(A \cap B) = \text{Prob}(B/A) \times \text{Prop}(A)$

P14. **If $Inc(\mathbf{A_1} \cap \mathbf{A_2} \cap \ldots \cap \mathbf{A_{n-1}}) = \tau_{\gamma(n-1)}$ with $\tau_{\gamma(n-1)} > \tau_0$**
then $Inc(\mathbf{A_1} \cap \mathbf{A_2} \cap \ldots \cap \mathbf{A_n}) = \tau_{\gamma n}$

$$\text{with } \gamma n = \begin{cases} \mathbf{MUL}_{\mathcal{L}}(Inc(\mathbf{A_1}), Inc(\mathbf{A_2/A_1})) & \text{if } n = 2 \\ \mathbf{MUL}_{\mathcal{L}}(\tau_{\gamma(n-1)}, Inc(\mathbf{A}n/\sum_{k=1}^{n} \mathbf{A_i}) & \text{if } n > 2 \end{cases}$$

In the classical probability theory, this property corresponds to the generalization of the composed probabilities formula. If $\text{Prob}(A_1 \cap A_2 \cap \ldots \cap A_n) > 0$ then
$\text{Prob}(A \cap B) = \text{Prob}(A_1) \times \text{Prob}(A_2/A_1) \times \ldots \times \text{Prob}(A_n/A_1 \cap A_2 \cap \ldots \cap A_{n-1})$

P15. **If A and B are two quasi–independent events and $Inc(\mathbf{B/A}) = \tau_\zeta$**
and $Inc(\mathbf{A}) = \tau_\alpha$ then $Inc(\neg\mathbf{B/A}) = \neg\tau_\zeta$
$\text{Prob}(\neg B/A) = 1 - \text{Prob}(B/A)$

P16. **If the events $\mathbf{B_1} \cup \mathbf{B_2}$ and A are quasi–independent and**
$Inc(\mathbf{B_1} \cup \mathbf{B_2/A}) = \tau_\zeta$ and $Inc(\mathbf{B_1} \cap \mathbf{B_2}) = \tau_0$
then $Inc(\mathbf{B_1} \cup \mathbf{B_2/A}) = \tau_\zeta = \mathbf{ADD}_{\mathcal{L}}(Inc(\mathbf{B_1/A}), Inc(\mathbf{B_2/A}))$
$\text{Prob}(B_1 \cap B_2/A) = \text{Prob}(B_1/A) + \text{Prob}(B_2/A)$ if $B_1 \cap B_2 = \emptyset$

P17. **If $Inc(\mathbf{A}) = \tau_\alpha$ in a complete system of two events**
then $Inc(\mathbf{B}) = \tau_\beta$ with $\tau_\beta = \mathbf{ADD}_{\mathcal{L}}(\mathbf{MUL}_{\mathcal{L}}(Inc(\mathbf{B/A}), Inc(\mathbf{A})),$
$\mathbf{MUL}_{\mathcal{L}}(Inc(\mathbf{B/\neg A}), Inc(\neg\mathbf{A})))$
Given a complete system of two events, we have:
$\text{Prob}(B) = \text{Prob}(B/A) \times \text{Prob}(A) + \text{Prob}(B/\neg A) \times \text{Prob}(\neg A)$

Notice that the following Bayes formulas are also expressed in the qualitative theory in [13]:
$\text{Prob}(B) = \sum_{i \in I} \text{Prob}(A_i)\text{Prob}(B/A_i)$

$\text{Prob}(A_k/B) = \text{Prob}(A_k)\,\text{Prob}(B/A_k)/\sum_{k=1}^{n}\text{Prob}(A_k)\,\text{Prob}(B/A_k)$

Using the above tools, the following rules are stated:

Symbolic logical implication rule. **If $Inc(\mathbf{A}) = \tau_\alpha$ and $Inc(\mathbf{B}) = \tau_\beta$**
and $Inc(\mathbf{A} \cap \mathbf{B}) = \tau_\gamma$ then $Inc(\mathbf{A} \supset \mathbf{B}) = Inc(\mathbf{A}) \rightarrow_{\mathbf{L}} Inc(\mathbf{A} \cap \mathbf{B})$
The conditional uncertainty $Inc(B/A)$ is comparable with the uncertainty implication $Inc(A \supset B)$

Proof: The evaluation of a qualitative uncertainty of "A implies B" denoted by $A \supset B$, whose premise A is τ_α–certain and whose conclusion B is τ_β–certain, preserves its meaning if and only if $Inc(A \cap B) \neq \tau_0$.

Let $Inc(A) = \tau_\alpha$, $Inc(B) = \tau_\beta$ and $Inc(A \cap B) = \tau_\gamma$ with $\tau_\gamma = \tau_{\alpha+\beta-(M-1)} > \tau_0$ be the uncertainty degrees of A, B and $A \cap B$. By definition of the logical implication,
$Inc(A \supset B) = Inc((A \cap B) \cup (B \backslash A \cap B))$ with $Inc(B \backslash A \cap B) = \max(\tau_{\beta-\gamma}, \tau_0) = \max(\tau_{\beta-\alpha-\beta+(M-1)}, \tau_0) = \max(\tau_{M-1-\alpha}, \tau_0) = \max(\neg\tau_\alpha, \tau_0) = \neg\tau_\alpha$
According to U3, $Inc(\neg A) = \neg\tau_\alpha$
Substituting $B \backslash A \cap B$ by $\neg A$ and as $A \cap B$ and $\neg A$ are incompatible:
$Inc(A \cap B \cup \neg A) = \text{ADD}_{\mathcal{L}}(\tau_\gamma, \neg\tau_\alpha) = \text{ADD}_{\mathcal{L}}(\neg\tau_\alpha, \tau_\gamma) = \min(\tau_{M-1-\alpha+\gamma}, \tau_{M-1})$
The last formula is the implication of Lukasiewicz for $(\tau_\alpha, \tau_\gamma)$ thus
$Inc(A \supset B) = Inc(A \cap B \cup \neg A) = Inc(A) \rightarrow_L Inc(A \cap B)$
Generalized modus ponens rule. **If $Inc(\mathbf{A} \supset \mathbf{B}) = \tau_\lambda$ and $Inc(\mathbf{A}) = \tau_\alpha$**
then $Inc(\mathbf{B}) = \tau_\beta$ with $\tau_\beta = \text{MUL}_{\mathcal{L}}(\tau_\alpha, \tau_\lambda) = \max(\tau_{\alpha+\lambda-(\mathbf{M-1})}, \tau_0)$

Proof: Let $Inc(A \cap B) = \tau_\gamma$ be the uncertainty degree of $A \cap B$
According to the symbolic logical implication rule, we have:
$Inc(A \supset B) = Inc(A) \rightarrow_L Inc(A \cap B) = \min(\tau_{M-1-\alpha+\gamma}, \tau_{M-1})$
Moreover, $\tau_\gamma \leq \tau_\beta \Longrightarrow \tau_\alpha \rightarrow_L \tau_\gamma \leq \tau_\alpha \rightarrow_L \tau_\beta$
And: $\tau_\alpha \rightarrow_L \tau_\beta = \tau_\gamma \Longrightarrow \tau_\lambda \leq \tau_\alpha \rightarrow_L \tau_\beta$
Let us consider the case where $\tau_\alpha > \tau_\beta$:
$\tau_\delta \leq \tau_{M-1-\alpha+\beta} \Longrightarrow \lambda \leq M - 1 - \alpha + \beta \Longrightarrow \beta \geq \alpha + \lambda - (M - 1)$
$\Longrightarrow \tau_\beta \geq \tau_{\alpha+\lambda-(M-1)}$ or $\tau_\beta \geq \max(\tau_{\alpha+\lambda-(M-1)}, \tau_0)$
For a good propagation of degrees we take: $\tau_\beta = \max(\tau_{\alpha+\lambda-(M-1)}, \tau_0)$

2.2.4 Consequences

Quantitative approach in the classical sense
Reminding the classical definition of conditional probability "the probability of B knowing A is equal to the ratio of $P(A \cap B) \neq 0$ and $P(A) \neq 0$", in other words:

$$P(B/A) = P(A \cap B)/P(A) \qquad (2.1)$$

Qualitative approach in the classical sense
The formulation of this approach is represented by U8:

$$Inc(B/A) = \text{DIV}_{\mathcal{L}}(Inc(A \cap B), Inc(A)) \text{ with } Inc(A \cap B) \neq \tau_0, Inc(A) \neq \tau_0 \qquad (2.2)$$

Qualitative approach in the inferential sense
The uncertainty of $A \supset B$ associated with the implication of Lukasiewicz can also be defined by:

$$Inc(A \supset B) = Inc(A) \rightarrow_L Inc(A \cap B) \text{ with } Inc(A \cap B) \neq \tau_0, Inc(A) \neq \tau_0 \qquad (2.3)$$

Expressions (2) and (3) are equivalent. The proof is easy to establish.

2.3 Symbolic Modification

Many authors such as Zadeh, Bouchon–Meunier, Akdag & al, Herrera & al, López de Mántaras & al, etc. have focused on the problem of data modification. Zadeh has introduced the notion of linguistic hedges [22] and other authors as Bouchon–Meunier has defined reinforcing and weakening fuzzy modifiers (defined thanks to fuzzy subsets) [23]. This notion of modifiers has been used in many fields. For example, López de Mántaras & Arcos have presented a system able to generate expressive music by means of fuzzy modification of parameters such as vibrato, dynamics, rubato... [24]. In [25], the modification between two fuzzy subsets is represented with a pair {linguistic term, translation} depending on a certain *hierarchy* (or precision level). The underlying framework is a scale of non uniformly distributed symbols. Several hierarchies are used to express the distribution of symbols in a fuzzy way.

In the many-valued logic context, another approach of data modification using uniformly distributed scales has been proposed [15]. Figure 2.1 shows an example with a scale of totally ordered degrees (in $\mathcal{L}_7$) and their linguistic counterpart.

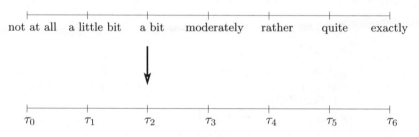

Fig. 2.1. Example of linguistic degrees and symbols

These scales refer to the measure theory. Usually the measure scales are divided into four classes: nominal, ordinal, interval and ratio scales [26]. A nominal scale does not express any values or relationships between variables (except equality). An ordinal scale is a scale on which data is shown simply in order of magnitude. These types of scale permit the measurement of degrees of difference, but not the specific amount of difference. An interval scale is a scale according to which the differences between values can be quantified in absolute but not relative terms and for which any zero is merely arbitrary: for instance, dates are measured on an interval scale since differences can be measured in years. A ratio scale permits the comparison of differences of values. It has a fixed zero value. Akdag & al scales are interval scales since there is no fixed zero value and differences of values can not always be compared.

In order to modify data, the degrees and/or the scales are modified. This is performed by dilation or erosion of the scales [15]. Thus this permits the obtention of modifiers that are functions allowing us to go from a degree to another. These *linguistic symbolic modifiers* have been generalized and formalized in [16]: the *generalized symbolic modifiers*.

2.3.1 Definitions

In this context a generalized symbolic modifier (GSM) corresponds to a semantic triplet of parameters: *radius, nature* (i.e. dilated, eroded or conserved) and *mode* (i.e. reinforcing, weakening or centring). The radius is denoted ρ with $\rho \in \mathbb{N}^*$. The more the radius, the more powerful the modifier. The nature is denoted by n and the mode by o.

Let us consider a symbolic degree $\tau_i \in \mathcal{L}_M$. The modified degree $\tau_{i'} \in \mathcal{L}_{M'}$ is computed by a function f thanks to a GSM m with a radius ρ:

$$\tau_{i'} = f_{m(\text{radius},\text{nature},\text{mode})}(\tau_i) = f_{m(\rho,n,o)}(\tau_i)$$

To simplify, another definition can be proposed:

A GSM m, denoted m_ρ, is a mapping where an initial degree τ_i (in the scale $\mathcal{L}_M$) is associated with a new degree $\tau_{i'}$ (the modified degree in the scale $\mathcal{L}_{M'}$) depending on a certain radius ρ:

$$m_\rho: \mathcal{L}_M \rightarrow \mathcal{L}_{M'}$$
$$\tau_i \mapsto \tau_{i'}$$

Let $\tau_i \in \mathcal{L}_M$ be a degree. The highest degree of the scale is denoted $\text{MAX}(\mathcal{L}_M) = M - 1$. The position of the degree in the scale is denoted $p(\tau_i) = i$.

A proportion or an intensity rate is associated with each linguistic degree on the considered scale: this rate corresponds to the ratio $\text{Prop}(\tau_i) = \frac{p(\tau_i)}{\text{MAX}(\mathcal{L}_M)}$.

According to ρ, the new associated degree $\tau_{i'}$ is more or less a neighbor of the initial degree τ_i. The higher the radius, the less neighboring the degrees. This way, three GSM families are defined: weakening, reinforcing and central ones. The first two families offer modifiers that weaken or reinforce the initial value $\text{Prop}(\tau_i)$, it means that $\text{Prop}(\tau_{i'}) > \text{Prop}(\tau_i)$ or $\text{Prop}(\tau_{i'}) < \text{Prop}(\tau_i)$. Modifiers from the third family don't alter the ratio but act as a zoom [16,27].

The three kinds of weakening GSMs are called EW, DW and CW for Eroded Weakening, Dilated Weakening and Conserved Weakening, depending on the treatment on the scale $\mathcal{L}_M$. In the same way reinforcing GSMs are called ER, DR and CR, with R for Reinforcing.

The definitions of these modifiers are summarized in Table 2.5.

Figure 2.2 shows examples of weakening modifiers and reinforcing. The radius that is used is each time equal to 1 to permit a better comparison.

Table 2.5. Summary of reinforcing and weakening generalized modifiers

Mode Nature	Weakening		Reinforcing	
Erosion	$\tau_{i'} = \max(\tau_0, \tau_{i-\rho})$ $\mathcal{L}_{M'} = \mathcal{L}_{\max(1,M-\rho)}$	$\mathbf{EW}(\rho)$	$\tau_{i'} = \tau_i$ $\mathcal{L}_{M'} = \mathcal{L}_{\max(i+1,M-\rho)}$	$\mathbf{ER}(\rho)$
			$\tau_{i'} = \min(\tau_{i+\rho}, \tau_{M-\rho-1})$ $\mathcal{L}_{M'} = \mathcal{L}_{\max(1,M-\rho)}$	$\mathbf{ER'}(\rho)$
Dilation	$\tau_{i'} = \tau_i$ $\mathcal{L}_{M'} = \mathcal{L}_{M+\rho}$	$\mathbf{DW}(\rho)$	$\tau_{i'} = \tau_{i+\rho}$ $\mathcal{L}_{M'} = \mathcal{L}_{M+\rho}$	$\mathbf{DR}(\rho)$
	$\tau_{i'} = \max(\tau_0, \tau_{i-\rho})$ $\mathcal{L}_{M'} = \mathcal{L}_{M+\rho}$	$\mathbf{DW'}(\rho)$		
Conservation	$\tau_{i'} = \max(\tau_0, \tau_{i-\rho})$ $\mathcal{L}_{M'} = \mathcal{L}_M$	$\mathbf{CW}(\rho)$	$\tau_{i'} = \min(\tau_{i+\rho}, \tau_{M-1})$ $\mathcal{L}_{M'} = \mathcal{L}_M$	$\mathbf{CR}(\rho)$

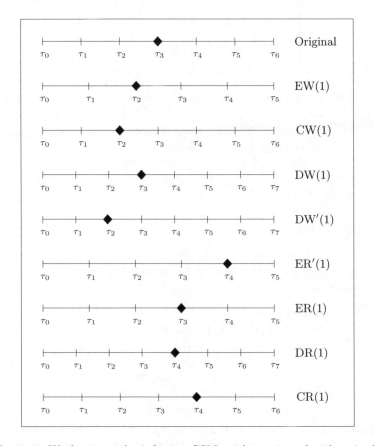

Fig. 2.2. Weakening and reinforcing GSMs with $\rho = 1$, and with τ_3 in $\mathcal{L}_7$

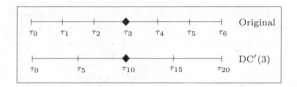

Fig. 2.3. Central GSM DC$'$ with $\rho = 3$, and with τ_3 in $\mathcal{L}_7$

The third GSMs family contains four kinds of central modifiers: EC, EC$'$ and DC, DC$'$ for Eroded Central and Dilated Central. As an example, Fig. 2.3 shows the behavior of DC$'(3)$ whose formal definition is the following:

$$
\text{DC}'(\rho) = \begin{cases} \tau_{i'} = \begin{cases} \tau_{i*(M\rho-1)\underline{\underline{M}}1} & \text{if } \tau_{i*(M\rho-1)\underline{\underline{M}}1} \in \mathcal{L}_{M'} {}^{,11} \\ \tau_{\lfloor i*(M\rho-1)\underline{\underline{M}}1 \rfloor} & \text{else (pessimistic)} \\ \tau_{\lfloor i*(M\rho-1)\underline{\underline{M}}1 \rfloor+1} & \text{else (optimistic)} \end{cases} \\ \mathcal{L}_{M'} = \mathcal{L}_{M\rho} \end{cases}
$$

In order to compare the GSMs, the ratio Prop is associated with each modifier. Indeed, if $\tau_i \overset{m_\rho}{\longmapsto} \tau_{i'}$ then $\text{Prop}(\tau_i) \overset{m_\rho}{\longmapsto} \text{Prop}(\tau_{i'})$. Moreover, if $\tau_{i'} = m_\rho(\tau_i)$ then $\text{Prop}(\tau_{i'}) = \text{Prop}(m_\rho(\tau_i)) = \text{Prop}(m_\rho)$ (misuse of writing).

For example, after applying a modifier m_ρ, if there are seven degrees ("not at all", "a little bit", "a bit", "moderately", "rather", "quite", "exactly") with $\tau_{i'}$ equal to "not at all", then the position of $\tau_{i'}$ is 0 within a scale of seven degrees and $\text{Prop}(m_\rho) = \frac{0}{6} = 0$.

2.3.2 Order Relation

Among the weakening and reinforcing GSMs, some of them are more powerful than some others [27]. To compare the GSMs, an order relation can be established between them.

Definition 2. *Let $m_{\rho,1}$ and $m_{\rho,2}$ be two modifiers with the same radius ρ. Let us denote $m_{\rho,1}(\tau_i) = \tau_{i'_1}$ with $\tau_{i'_1} \in \mathcal{L}_{M'_1}$ and $m_{\rho,2}(\tau_i) = \tau_{i'_2}$ with $\tau_{i'_2} \in \mathcal{L}_{M'_2}$*
$\text{Prop}(m_{\rho,1}) = \frac{p(\tau_{i'_1})}{\text{MAX}(\mathcal{L}_{M'_1})}$ is comparable with $\text{Prop}(m_{\rho,2}) = \frac{p(\tau_{i'_2})}{\text{MAX}(\mathcal{L}_{M'_2})}$ if and
only if, for any M, $\begin{cases} \text{Prop}(m_{\rho,1}) \le \text{Prop}(m_{\rho,2}) \;\; \forall \tau_i \in \mathcal{L}_M \\ \text{or} \\ \text{Prop}(m_{\rho,2}) \le \text{Prop}(m_{\rho,1}) \;\; \forall \tau_i \in \mathcal{L}_M \end{cases}$

Definition 3. *Two modifiers $m_{\rho,1}$ and $m_{\rho,2}$ entail an order relation if and only if $\text{Prop}(m_{\rho,1})$ is comparable with $\text{Prop}(m_{\rho,2})$, for any given $\tau_i \in \mathcal{L}_M$. Formally, the relation $\preceq$ is defined as follows:*

For any M, $m_{\rho,1} \preceq m_{\rho,2} \Leftrightarrow \text{Prop}(m_{\rho,1}) \le \text{Prop}(m_{\rho,2}) \;\; \forall \, \tau_i \in \mathcal{L}_M$

---

[11] In other words, if $i * (M\rho - 1)\underline{\underline{M}}1 \in \mathbb{N}^* \smallsetminus \{1\}$.

It is to note that if a pair of modifiers $(m_{\rho,1}, m_{\rho,2})$ are in relation to each other, the comparison between their intensity rates is possible obviously because these intensity rates are rational numbers but particularly because the unit is the same, for all $\tau_i \in \mathcal{L}_M$. Indeed, the degrees are uniformly distributed on the scales. Furthermore, it is easy to see that the binary relation $\preceq$ over the generalized modifiers is a partial order relation.

By comparing the generalized modifiers in pairs, it is possible to establish a partial order relation between them that can be expressed through a lattice (cf. Fig. 2.4). The relation is only partial because some modifiers can not be compared with some others.

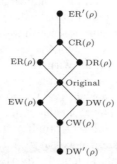

Fig. 2.4. Lattice for the relation $\preceq$

Notice that the central GSMs don't appear in the lattice since they don't change the proportions Prop.

2.3.3 Finite and Infinite Modifiers

As the radius can be instanciated by any positive integer, the GSM may modify a bit, a lot or enormously the initial value. Following this idea, an interesting point is to see if the limits of the scale can be reached. In this perspective, notions of finite and infinite modifiers are defined.

Definition 4. *An infinite modifier m_ρ is defined as follows:*

$$m_\rho \text{ is an infinite modifier} \Leftrightarrow \begin{cases} (\forall \rho \in \mathbb{N}^*, \; Prop(m_{\rho+1}) > Prop(m_\rho)) \\ or \\ (\forall \rho \in \mathbb{N}^*, \; Prop(m_{\rho+1}) < Prop(m_\rho)) \end{cases}$$

This means that the modifier will always have an effect on the initial value. For example, $DW(\rho)$ and $DR(\rho)$ can modify the initial value towards infinity.

Definition 5. *A finite modifier* m_ρ *is defined as follows:*

$$m_\rho \text{ is a finite modifier} \Leftrightarrow \begin{cases} \exists \rho \in \mathbb{N}^* \text{ such as } \forall \rho' \in \mathbb{N}^* \text{ with } \rho' > \rho \\ Prop(m_{\rho'}) = Prop(m_\rho) \end{cases}$$

This means that, starting from a certain rank, the modifier has no effect on the initial value.

2.3.4 Composition

Another interesting result concerns the *composition* (in the mathematical sense) of the GSMs [28]. For example, composing a modifier ER with a modifier DR consists in applying (on an initial value) first a modifier DR and then a modifier ER. Two kinds of composition have to be distinguished: *homogeneous* and *heterogeneous* ones. Homogeneous compositions are compositions of modifiers from the same family with the same nature, same mode, same name but not necessarily the same radius (for example, a DW modifier with a DW modifier, or DW' modifier with a DW' modifier...). Any other form of compositions is said to be heterogeneous: it means that heterogeneous compositions imply GSMs from different families. These compositions permit to reach any degree on any scale if necessary. Moreover, the linguistic counterpart can be expressed in combinations of adverbs, such as, for example, "very very" corresponding to $CR(\rho) \circ CR(\rho)$. Thus, the following theorem can be easily proved [28]:

Theorem 1. *The result of the composition of generalized symbolic modifiers is also a generalized symbolic modifier: when composing n generalized symbolic modifiers (of any kinds), a valid pair degree/scale is always obtained.*

Concerning the homogeneous compositions, two other important theorems have been proved also in [28]:

Theorem 2. *If* m_{ρ_1} *is any weakening or reinforcing GSM with a radius* ρ_1, *and if* m_{ρ_2} *is any GSM of the same family than* m_{ρ_1} *with a radius* $\rho_2, \ldots$ *and if* m_{ρ_n} *is any GSM of the same family than* m_{ρ_1} *with a radius* ρ_n, *then* $m_{\rho_s} = m_{\rho_1} \circ m_{\rho_2} \circ \ldots \circ m_{\rho_n}$ *is a GSM of the same mode than* m_{ρ_1}, *but with a radius* ρ_s *equal to the sum of the radiuses, i.e.* $\rho_1 + \rho_2 + \ldots + \rho_n$.

For example, $\forall \rho_1, \rho_2, \ldots \rho_n \in \mathbb{N}^*$, $DR(\rho_1) \circ DR(\rho_2) \circ \ldots \circ DR(\rho_n) = DR(\rho_1 + \rho_2 + \ldots + \rho_n)$.

A similar theorem for central modifiers is also defined in [28].

Theorem 3. *If* m_{ρ_1} *is a GSM DC with a radius* ρ_1, *and if* m_{ρ_2} *is a GSM DC with a radius* $\rho_2, \ldots$ *and if* m_{ρ_n} *is a GSM DC with a radius* ρ_n, *then* m_{ρ_p} *is a GSM DC with a radius* ρ_p *equal to the product of the radiuses, i.e.* $\rho_1 \rho_2 \ldots \rho_n$.

In other words, $\forall \rho_1, \rho_2 \ldots \rho_n \in \mathbb{N}^*$, $DC(\rho_1) \circ DC(\rho_2) \circ \ldots \circ DC(\rho_n) = DC(\rho_1 \rho_2 \ldots \rho_n)$.

2.4 Symbolic Combination: A Median Operator

Another important problem is the combination of symbolic data under the considered many-valued logic. The four principal operators: addition, subtraction, division and multiplication have been previously defined. Many other kinds of combination operators exist: for example, arithmetic mean, geometric mean, medians... But the last operators (or aggregators) usually act on numbers and not symbols in the general sense. In some other cases, aggregators are not very accurate. For example, let us consider the median definition with weighted elements:

Definition 6. *Considering M distinct elements $x_0, x_1, \ldots, x_{M-1}$ with positive weights $w_0, w_1, \ldots, w_{M-1}$ such as $\sum_{i=0}^{M-1} w_i = 1$, the weighted median ("pessimistic"[12] or lower bound) is the element x_k satisfying : $\sum_{x_i < x_k} w_i < 1/2$ and $\sum_{x_i > x_k} w_i \leq 1/2$.*

In the above definition the M elements have to assume an order relation [29].

This median may be unaccurate when the elements are very heterogenous: Let $(0,1,3,100,110,111)$ be the given set. Each element has the same weight than the others: $1/6$. Depending on the chosen definition, the median is 3 (pessimistic) or 100 (optimistic).

Another possibility to combine a set of symbolic data is to define a center of gravity. A typical use of such an operator is the example of an opinion poll. The opinion poll is composed of a question and a set of possible answers. Each answer is a label (linguistic expression) and the set of answers is totally ordered. Each label is assigned to a weight corresponding to the percentage of persons that have chosen this answer. For example:

– Question : What do you think about music courses at school?

   ▷ useless   → 5 %
   ▷ boring   → 3 %
   ▷ moderately interesting → 19 %
   ▷ interesting   → 65 %
   ▷ exciting   → 8 %

The center of gravity here can be represented by a symbolic median. The operator is based on the principle of the classic median.

---

[12] The median is said "optimistic" when the inequalities are changed (i.e. respectively $\leq$ and $<$).

2.4.1 Definitions

Knowing the M possible (weighted) answers, let us consider a tree with a root and its M leaves. The weight of each answer is associated with the corresponding leaf, and a leaf may have a zero-weight. Conventionnally, the answers are ordered on the tree, with the weakest at the top. $\tau_{i,M-1}$ is the ith possible initial answer.

Graphically, two trees with M leaves are actually considered: the first one permits to express the titles $\tau_{i,M-1}$ of the possible answers and the second one permits to express the weights of the answers (cf. Fig. 2.5).

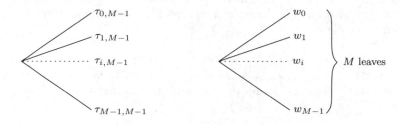

Fig. 2.5. Representation with two trees

Let T_1 and T_2 be two trees with M leaves associated with an answer $\tau_{i,M-1}$ and a weight w_i, with $w_i \in [0,1]$ and $\sum_{i=0}^{M-1} w_i = 1$.

Let $\tilde{\mathcal{L}}_M$ be the set of the initial possible answers with $\tilde{\mathcal{L}}_M = \{\tau_{0,M-1}, \tau_{1,M-1}, \ldots, \tau_{M-1,M-1}\}$.

An answer whose label (identification) is id and whose weight is w is denoted $\tau_{id,M-1}^w$.

The notation $(\tau_{id,M-1}^w)$ that takes into account weights permits to lighten the graphic representation: instead of using two trees, only one is enough (cf. Fig. 2.6).

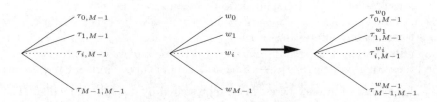

Fig. 2.6. Representation with one tree

Definition 7. *Let* $\tilde{\mathcal{L}}_M^w = \{\tau_{0,M-1}^{w_0}, \tau_{1,M-1}^{w_1}, \dots, \tau_{M-1,M-1}^{w_{M-1}}\}$ *be a set of weighted ordered* answers*. The symbolic weighted median* $\mathcal{M} \in \tilde{\mathcal{L}}_{M'}$ *is defined as follows:*

$$\mathcal{M}: \quad \tilde{\mathcal{L}}_M^w \to \tilde{\mathcal{L}}_{M'}^{w'}$$

$$\left(\tau_{0,M-1}^{w_0}, \tau_{1,M-1}^{w_1}, \dots, \tau_{M-1,M-1}^{w_{M-1}}\right) \mapsto \mathcal{M}(\tau_{0,M-1}^{w_0}, \tau_{1,M-1}^{w_1}, \dots, \tau_{M-1,M-1}^{w_{M-1}})$$

$$= \tau_{i,M'-1}^{w_i'}, \text{ the leaf of the sub-tree (whose}$$
$$\text{root is } \tau_{i,M-1}^{w_i}) \text{ that is constructed}$$
$$\text{according to the algorithm below and}$$
$$\text{such that:} \quad \left| \sum_{p=0}^{i-1} w_p' - \sum_{p=i+1}^{M'-1} w_p' \right| < \varepsilon$$

Leaf $\tau_{i,M'-1}^{w_i'}$ doesn't necessarily belong to the initial tree, i.e. leaf $\tau_{i,M'-1}^{w_i'}$ doesn't necessarily belong to the set $\{\tau_{0,M-1}^{w_0}, \tau_{1,M-1}^{w_1}, \dots, \tau_{M-1,M-1}^{w_{M-1}}\}$, but may be a *new* leaf constructed from the initial tree and according to the following algorithm:

Algorithm 2.1 Symbolic Weighted Median Construction

The principle is to look for a certain leaf l. l is the leaf for which the sum of the weights of the leaves that precede l is equal to the sum of the weights of the leaves that come next l in the tree. In practice, the obtention of the equality or near equality (thanks to ε) of the sums is rare (not frequent). Therefore, if such a leaf can't be found directly, a certain procedure has to be performed in order to "refine" the trees — the "refinement" consists in the subdivision of the leaves (they become nodes), i.e. the creation of children for each node. This way, the obtention of the equality or near equality of the sums of weights become possible.

- If a leaf for which the sum of its weight and the weights of the preceding leaves is exactly equal to $1/2$, is found, then $M - 1$ leaves, with a zero-weight, have to be added to the tree. The $M - 1$ new leaves must come in between the existing ones. Then the sums are computed again (cf. Fig. 2.7).
- Else the algorithm looks for the first leaf whose weight is denoted w_k and such that $\sum_{i=0}^{k} w_i > 1/2$. Let $w_\sigma = (\sum_{i=0}^{k} w_i) - \frac{1}{2}$. Then each leaf (node) has to be subdivided into η leaves, with $\eta = \lceil \frac{w_k}{w_\sigma} \rceil$. The weight assigned to each new leaf is the weight of the parent node divided by η (cf. Fig. 2.8).
- These conditional tests may be reiterated until obtaining the near equality of the sums.

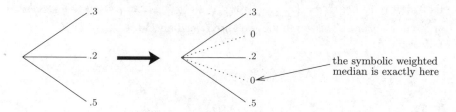

Fig. 2.7. Median: example 1

It is to notice that, the less ε, the more precise the median.

After obtaining the complete tree, the symbolic weighted median (SWM) $\mathcal{M}$ may be denoted in the following way:

$$\mathcal{M}\left(\tau_{0,M'-1}^{w_0'}, \tau_{1,M'-1}^{w_1'}, \ldots, \tau_{M'-1,M'-1}^{w_{M'-1}}\right)$$

$$= \left\{ \tau_{i,M'-1}^{w_i'} \text{ t.q } \exists \varepsilon > 0, \left| \sum_{p=0}^{i-1} w_p' - \sum_{p=i+1}^{M'-1} w_p' \right| < \varepsilon \right\} = t\left(\tau_{k,M-1}^{w_k}\right)$$

with $\tau_{0,M'-1}^{w_0'}, \tau_{1,M'-1}^{w_1'}, \ldots, \tau_{M'-1,M'-1}^{w_{M'-1}}$ corresponding to the M' leaves (with $w_0', w_1', \ldots, w_{M'-1}'$ their weights) of the complete tree, where $\tau_{k,M-1}^{w_k}$ is the parent node – from the initial tree – of $\tau_{i,M'-1}^{w_i'}$.

In this expression, the aggregated answer is defined as a transformation t of one of the initial answers $\tau_{k,M-1}^{w_k}$. As it will be proven in theorem 4, this transformation t is a generalized symbolic modifier m. According to Figs. 2.7 and 2.8, the symbolic weighted median is one of the M' leaves or, in some limit cases, one of the M leaves.

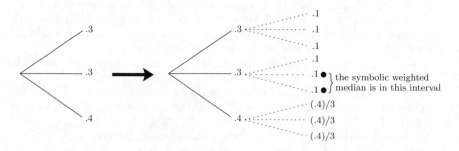

Fig. 2.8. Median: example 2

leaf k
symbolic weighted median
central GSM
reinforcing GSM

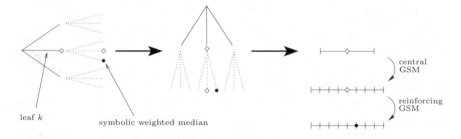

Fig. 2.9. The symbolic weighted median and the GSMs

2.4.2 Expressing the Median

The initial tree is now considered as a scale with M linguistic degrees. During the subdivision step, the number of degrees of the "scale" increases, thanks to an action like a zoom: the scale is dilated. Thus, leaf k (in the algorithm) corresponds to degree $\tau_{k,M-1}^{w_k}$ in the scale $\tilde{\mathcal{L}}_M^w$. Therefore, tree subdivisions are similar to the application of central GSMs and the median is obtain thanks to jumps similar to weakening or reinforcing GSMs (cf. Fig. 2.9).

Consequently, each leaf of the complete tree is a modifier (weakening, reinforcing or central) — or a composition of modifiers — of its parent node that belongs to the initial tree (cf. Fig. 2.10).

Thus, as soon as the SWM is computed, the GSM(s) is (are) obtained. There is a correspondence between the SWM and the GSMs that is defined as follows:

Definition 8. *Let $\tau_{i,M'-1}^{w'_i}$ be the leaf corresponding to the SWM and $\tau_{k,M-1}^{w_k}$ be the parent node of the SWM. Let A and B be two rational numbers:*

$$A = \frac{i}{M'-1} \quad \textit{and} \quad B = \frac{k}{M-1}$$

$\tau_{0,2}^{w_0}$
$\tau_{1,2}^{w_1}$
$\tau_{2,2}^{w_2}$
central GSM with regard to node $\tau_{0,2}^{w_0}$
reinforcing GSM with regard to node $\tau_{0,2}^{w_0}$
reinforcing GSM with regard to node $\tau_{0,2}^{w_0}$
weakening GSM with regard to node $\tau_{1,2}^{w_1}$
central GSM with regard to node $\tau_{1,2}^{w_1}$
reinforcing GSM with regard to node $\tau_{1,2}^{w_1}$
weakening GSM with regard to node $\tau_{2,2}^{w_2}$
weakening GSM with regard to node $\tau_{2,2}^{w_2}$
central GSM with regard to node $\tau_{2,2}^{w_2}$

Fig. 2.10. GSMs associated to the SWM

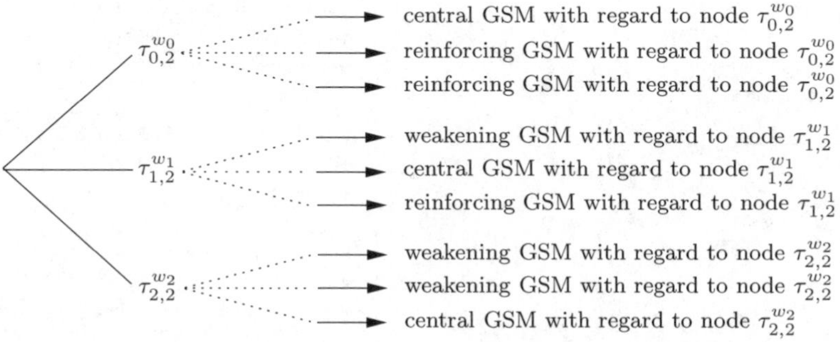

In some cases, the computation of B is not easy: in Fig. 2.7, for example, the parent node of the SWM is undefined. However, either of the following two options may be chosen: the parent node is the preceding node of the SWM (pessimistic choice) or the parent node is the next one (optimistic choice).

In Fig. 2.7, the pessimistic choice gives $B = 1/2$ (and $A = 3/5$) while the optimistic choice gives $B = 2/2$.

If no tree subdivision has been performed, the SWM is exactly one of the initial answers and then no GSM is associated to the median.

If, at least, one subdivision has been performed, three cases are possible, depending on the value of the difference $A - B$.

1. $A - B = 0 \Rightarrow$ Central GSMs. In this case, the corresponding GSMs are $DC'(\rho)$ that increase the precision: it corresponds to a change of scale, i.e. to a subdivision of the tree. ρ is equal to η. The linguistic term that can be associated is "PRECISELY". If several subdivisions are performed, several $DC'(\rho)$ are used in a composition. *According to Theorem 1, the composition of n GSMs is still a GSM.*
2. $A - B > 0 \Rightarrow$ Reinforcing GSMs. In this case, the corresponding GSMs are $DC'(\rho_1)$ composed with $CR(\rho_2)$. $CR(\rho)$ permits to increase the value without modifying the scale. ρ_1 is equal to η and ρ_2 is equal to $\lfloor |A - B| * (M' - 1) \rfloor$ (cf. example in Fig. 2.11).
3. $A - B < 0 \Rightarrow$ Weakening GSMs. In this case, the corresponding GSMs are $DC'(\rho_1)$ composed with $CW(\rho_2)$. $CW(\rho)$ permits to increase the value without modifying the scale. ρ_1 is equal to η and ρ_2 is equal to $\lfloor |A - B| * (M' - 1) \rfloor$ (cf. example in Fig. 2.11).

In Fig. 2.11, three different cases are shown:

First case. The subdivision with $\eta = 3$ corresponds to $DC'(3)$ and the second step corresponds to $CR(\rho)$ with $\rho = \lfloor |A - B|(M' - 1) \rfloor = \lfloor |5/8 - 1/2|(9 - 1) \rfloor = 1$. The transformation t is shown in Fig. 2.12 and can be easily expressed linguistically.

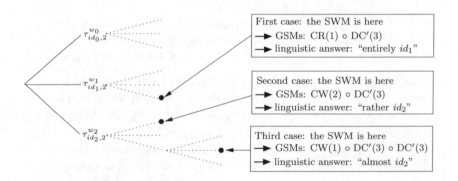

Fig. 2.11. Examples of cases where $A - B > 0$ and where $A - B < 0$

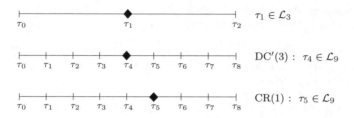

$\tau_1 \in \mathcal{L}_3$

$DC'(3) : \tau_4 \in \mathcal{L}_9$

$CR(1) : \tau_5 \in \mathcal{L}_9$

Fig. 2.12. Composition of GSMs $DC'(\rho)$ and $CR(\rho)$

With an ordered alphabet composed of x linguistic terms — $x/2$ reinforcing terms and $x/2$ weakening terms — such as: {very few; few; rather; almost; entirely; very; a lot; extremely}, $DC'(3)$ is translated by "precisely" and $CR(1)$ by the first (because $\rho = 1$) reinforcing term, i.e. "entirely". Theoretically, the combination of these linguistic answers (id_0, id_1 and id_2) would give "precisely entirely id_1". Practically, words can generally not be compound like mathematical tools such as GSMs: one should omit the linguistic term given by GSM $DC'(\rho)$, except when the median is obtained only thanks to a $DC'(\rho)$. The combination gives: "entirely id_1".

Second case. The subdivision with $\eta = 3$ corresponds to $DC'(3)$ and the second step corresponds to $CW(\rho)$ with $\rho = \lfloor |A - B|(M' - 1) \rfloor = \lfloor |6/8 - 2/2|(9 - 1) \rfloor = 2$. The transformation t is shown in Fig. 2.13 and can be easily expressed linguistically.

With the same alphabet, the combination gives: "rather id_2" because the second ($\rho = 2$) weakening term is "rather" while the first weakening one is "almost".

Third case. Following the same reasoning, the combination here gives: "almost id_2". In this case, it is to notice that the linguistic counterpart of the SWM is more approximative than its mathematical expression through GSMs. However, according to the alphabet, the linguistic expression can be more precise.

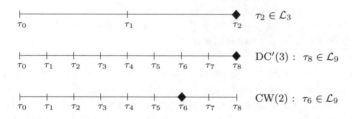

$\tau_2 \in \mathcal{L}_3$

$DC'(3) : \tau_8 \in \mathcal{L}_9$

$CW(2) : \tau_6 \in \mathcal{L}_9$

Fig. 2.13. Composition of GSMs $DC'(\rho)$ and $CW(\rho)$

So, it has been seen that the symbolic weighted median depends on GSMs. Thus, the final result is the following:

Theorem 4. *Let* $\tilde{\mathcal{L}}_M^w = \{\tau_{0,M-1}^{w_0}, \tau_{1,M-1}^{w_1}, \ldots, \tau_{M-1,M-1}^{w_{M-1}}\}$ *be a set of weighted ordered answers; let* m *be a generalized symbolic modifier; let* $\tau_{k,M-1}^{w_k}$ *be the parent node of the SWM:*

$$\mathcal{M}\left(\tau_{0,M-1}^{w_0}, \tau_{1,M-1}^{w_1}, \ldots, \tau_{M-1,M-1}^{w_{M-1}}\right) = m\left(\tau_{k,M-1}^{w_k}\right) = \tau_{j,M'-1}$$

with M' *the number of leaves of the complete tree and with* $\tau_{j,M'-1} \in \mathcal{L}_M'$.

It is to notice that the SWM $(\tau_{j,M'-1})$ has no associated weight, as it is the case for the classical aggregators.

The symbolic weighted median, as other aggregators, satisfies also the following properties that are proven in [28]:

- Identity, monotonicity, symmetry, idempotence, compensation.
- Boundary conditions: this property means that the aggregation, p times, of the lowest element of the tree is *this* element itself. Moreover, the aggregation, p times, of the highest element of the tree is *this* element itself.
- Continuity (adapted in this case of discrete elements): when the elements of the tree are slightly changed, the aggregator gives a result slightly different from the original result.
- Counterbalancement: adding weight on leaves placed above the SWM on the tree, will decrease the final result. And, reciproquely, adding weight on leaves placed below the SWM on the tree, will increase the final result.

2.5 Conclusion

The presented work provides efficient tools to manipulate, modify and merge symbolic data under an uncertain environment, and, more generally, in a qualitative framework. In this context data can be words, concepts, symbols (in the general sense)... and they are usually expressed under a linguistic form. The use of qualitative values is necessary in areas where the exact numerical values associated with a fact are unavailable or unknown by experts. The experts can, in general, only propose intervals of values, each interval being represented by a symbolic value.

The approach proposed in Sect. 2 for the representation and management of qualitative uncertainty, is conformable with the intuition, especially through the four symbolic arithmetic operations. Morevover, with its axiomatic of symbolic probabilities, this model constitutes an interesting alternative to the classical probability theory. Advanced theorems of this axiomatic permit to manipulate Bayesian networks and to manage uncertainties in Knowledge-Based Systems. In other terms, they allow us to reason under uncertainty or imprecision.

In the many-valued logic reminded in this chapter, data modification is performed thanks to degrees and scales: indeed information is converted into sets of ordered degrees attached to variables. The tools that perform the modifications (*generalized linguistic modifiers*: GSMs) permit to handle linguistic expressions such as "much more than...". They are characterized through a formal mathematical definition on the scales and degrees.

Depending on the cases, degrees can be balanced or not: weights are allocated to each degree in order to render the importance of a degree compared to the others. The combination of such weighted degrees is interesting in many cases, for instance when a linguistic abstract is needed. The *symbolic weighted median* (SWM) permits to obtain a representative degree, or a center of gravity of the set. It can be seen as an aggregator with the properties of a median.

The generalized symbolic modifiers described in this study permit the modification of symbolic values. Three kinds of GSM exist: weakening, reinforcing and centring ones. These modification tools entail an order relation and are presented in a lattice. This way, it is easy to choose (according to its strength) the GSM(s) we need for a given application.

The compositions of GSMs permit to express linguistic assertions composed of more than one adverb. Moreover, a certain composition of GSMs can express the result of an aggregation: the symbolic weighted median gives the emergent element of a set of weighted degrees. The median algorithm underlies the construction of the emergent element, and the aggregator assumes good mathematical properties usually desired for such an aggregator.

When computing the SWM with modifiers (the GSMs), an association is implicitly done between modifiers and aggregation. Actually this kind of aggregation can be seen as finding a modification of an element from the initial set.

The qualitative tools (GSMs and SWM) presented in this document may be useful in many practical applications. For instance, a suitable application domain is colorimetry [30–32]. Indeed one may need to know the color located between two given colors knowing that the sought color is more or less close to one of them. Moreover one may need to express this color with a linguistic set composed of a qualifier and a modifier (e.g. "a green *a bit less dark*"). In this kind of applications the scales and degrees are respectively the components of the color and their value in the considered space, for example, in RGB (Red–Green–Blue) space or HLS (Hue–Lightness–Saturation) space. Hence the GSMs help to find the desired color according to a linguistic expression. As for the median (SWM) colorimetry also constitutes a good application domain because one may need to express the color that satisfies most people in an opinion poll, for instance. Weights may correspond to the number of persons or to the importance of the persons. In this case colors with their associated weights (to be coherent, the colors have to be visually close to each other, such as "yellow and green") are aggregated thanks to the SWM. The composition of GSMs that permits to build the SWM corresponds to a linguistic expression, such as "mustard much deeper", for example.

References

1. H. Akdag, M. D. Glas, and D. Pacholczyk, "A qualitative theory of uncertainty," *Fundamenta Informaticae*, vol. 17, no. 4, pp. 333–362, 1992.
2. H. Akdag and F. Khoukhi, "Une approche logico-symbolique pour le traitement des connaissances nuancées," in *proceedings of the fifth IPMU*, (Paris), 1994.
3. L. Godo, R. López de Mántaras, C. Sierra, and A. Verdaguer, "Milord: The architecture and the management of linguistically expressed uncertainty," *International Journal of Intelligent Systems*, vol. 4, no. 4, pp. 471–501, 1989.
4. F. Nef, *La logique du langage naturel.* Hermès, 1989.
5. L. SOMBE, "Inférences non classiques en intelligence artificielle: Ebauches de comparaison sur un exemple," *P.R.C.-G.R.E.C.O "Intelligence Artificielle"*, 1988. Hermès.
6. M. De Glas, "Knowledge representation in a fuzzy setting," report 89–48, LAFORIA, University of Paris VI, 1989.
7. F. Khoukhi, *Approche logico-symbolique du traitement des connaissances incertaines et imprécises dans les systèmes à base de connaissances.* PhD thesis, University of Reims, 1996.
8. H. Seridi, F. Bannay-Dupin De St-Cyr, and H. Akdag, "Qualitative operators for dealing with uncertainty," in *proceeding of Fuzzy-Neuro Systems'98* (C. Freksa, ed.), (München), pp. 202–209, 1998.
9. M. De Glas, "Representation of Lukasiewicz' many-valued algebras ; the atomic case," *Fuzzy Sets and Systems*, vol. 14, 1987.
10. L. A. Zadeh, "The concept of linguistic variable and its application in approximate reasoning," *Information Science (I, II, III)*, vol. 8, no. 9, 1975.
11. R. Aleliunas, "A summary of new normative theory of probabilistic logic," *Uncertainty in Artificial Intelligence*, vol. 4, 1990. Elsevier Publication Science, North Holland.
12. A. Darwiche and M. Ginsberg, "A symbolic generalization of probability theory," in *proceedings of the American Association for Artificial Intelligence*, (San Jose, California), 1992.
13. H. Seridi and H. Akdag, "Approximate reasoning for processing uncertainty," *Journal of Advanced Computational Intelligence*, vol. 5, no. 2, pp. 108–116, 2001. Fuji technology Press.
14. H. Seridi, H. Akdag, and A. Meddour, "Une approche qualitative sur le traitement de l'incertain: Application au système expert," *Sciences et Technologies*, vol. 19, pp. 13–19, 2003. Constantine.
15. H. Akdag, N. Mellouli, and A. Borgi, "A symbolic approach of linguistic modifiers," in *Information Processing and Management of Uncertainty in Knowledge-Based Systems (IPMU), Madrid*, (Madrid), pp. 1713–1719, 2000.
16. H. Akdag, I. Truck, A. Borgi, and N. Mellouli, "Linguistic modifiers in a symbolic framework," *International Journal of Uncertainty, Fuzziness and Knowledge-Based Systems*, vol. 9 (Supplement), pp. 49–61, 2001.
17. M. Seban, *Modèles théoriques en reconnaissance de forme et architecture hybride pour machine perspective.* PhD thesis, University of Lyon 1, 1996.
18. D. Pacholczyk and J. M. Pacholczyk, "Traitement symbolique des informations incertaines," in *proceedings of RFIA*, (Nantes), 1996.
19. R. López de Mántaras, *Approximate Reasoning Models.* Ellis Horwood Series in Artificial Intelligence, 1990.

20. H. Seridi and H. Akdag, "A qualitative approach for processing uncertainty," *Uncertainty in Intelligent and Information Systems*, vol. 20, pp. 46–57, 2000. World Scientific, Advances in Fuzzy Systems-Applications and Theory.
21. B. Bouchon-Meunier, *La logique floue et ses applications.* Addison–Wesley, 1995.
22. L. A. Zadeh, "A fuzzy-set-theoretic interpretation of linguistic hedges," *Journal of Cybernetics*, vol. 2(3), pp. 4–34, 1972.
23. B. Bouchon-Meunier, "Stability of linguistic modifiers compatible with a fuzzy logic," *Uncertainty in Intelligent Systems, Lecture Notes in Computer Science, Springer–Verlag*, vol. 313, 1988.
24. R. López de Mántaras and J. L. Arcos, "AI and music: From composition to expressive performances," *AI Magazine*, vol. 23(3), pp. 43–57, 2002.
25. F. Herrera and L. Martínez, "A model based on linguistic 2-tuples for dealing with multigranularity hierarchical linguistic contexts in multiexpert decision-making," *IEEE Transactions on Systems, Man and Cybernetics. Part B: Cybernetics*, vol. 31, no. 2, pp. 227–234, 2001.
26. M. Grabisch, *Evaluation subjective, Méthodes, Applications et Enjeux*, ch. IV. ECRIN, 1997.
27. I. Truck, A. Borgi, and H. Akdag, "Generalized modifiers as an interval scale: towards adaptive colorimetric alterations," in *The 8th Iberoamerican Conference on Artificial Intelligence, IBERAMIA 2002* (Springer-Verlag, ed.), (Sevilla, Spain), pp. 111–120, 2002.
28. I. Truck, *Approches symbolique et floue des modificateurs linguistiques et leur lien avec l'agrégation.* PhD thesis, University of Reims, 2002.
29. R. R. Yager, "On weighted median aggregation," *International Journal of Uncertainty, Fuzzyness and Knowledge-Based Systems*, vol. 2, pp. 101–113, 1994.
30. A. Aït Younes, I. Truck, H. Akdag, and Y. Remion, "Image classification according to the dominant colour," in *6th International Conference on Enterprise Information Systems, ICEIS'04*, (Porto, Portugal), pp. 505–510, 2004.
31. A. Aït Younes, I. Truck, H. Akdag, and Y. Remion, "Image retrieval using linguistic expressions of colors," in *6th International Fuzzy Logic and Intelligent technologies in Nuclear Science on Applied Computational Intelligence (FLINS 2004)*, (Blankenberge, Belgium), 2004. to appear.
32. I. Truck, H. Akdag, and A. Borgi, "A symbolic approach for colorimetric alterations," in *2nd International Conference in Fuzzy Logic and Technology (EUSFLAT 2001)*, (Leicester, England), pp. 105–108, 2001.

Adaptation of Fuzzy Inference System Using Neural Learning

A. Abraham

Computer Science Department, Oklahoma State University, USA
ajith.abraham@ieee.org, http://ajith.softcomputing.net

The integration of neural networks and fuzzy inference systems could be formulated into three main categories: cooperative, concurrent and integrated neuro-fuzzy models. We present three different types of cooperative neuro-fuzzy models namely fuzzy associative memories, fuzzy rule extraction using self-organizing maps and systems capable of learning fuzzy set parameters. Different Mamdani and Takagi-Sugeno type integrated neuro-fuzzy systems are further introduced with a focus on some of the salient features and advantages of the different types of integrated neuro-fuzzy models that have been evolved during the last decade. Some discussions and conclusions are also provided towards the end of the chapter.

3.1 Introduction

Hayashi et al. [21] showed that a feedforward neural network could approximate any fuzzy rule based system and any feedforward neural network may be approximated by a rule based fuzzy inference system [31]. Fusion of Artificial Neural Networks (ANN) and Fuzzy Inference Systems (FIS) have attracted the growing interest of researchers in various scientific and engineering areas due to the growing need of adaptive intelligent systems to solve the real world problems [5,6,10–13,17,19,20,22,33,37]. A neural network learns from scratch by adjusting the interconnections between layers. Fuzzy inference system is a popular computing framework based on the concept of fuzzy set theory, fuzzy *if-then* rules, and fuzzy reasoning. The advantages of a combination of neural networks and fuzzy inference systems are obvious [12, 32]. An analysis reveals that the drawbacks pertaining to these approaches seem complementary and therefore it is natural to consider building an integrated system combining the concepts [37]. While the learning capability is an advantage from the viewpoint of fuzzy inference system, the automatic formation of linguistic rule base will be advantage from the viewpoint of neural network. There are several works related to the integration of neural networks and fuzzy inference systems [1–3, 15, 17, 23, 28, 30, 32, 34, 43, 45, 49, 52].

3.2 Cooperative Neuro-Fuzzy Systems

In the simplest way, a cooperative model can be considered as a preprocessor wherein artificial neural network (ANN) learning mechanism determines the fuzzy inference system (FIS) membership functions or fuzzy rules from the training data. Once the FIS parameters are determined, ANN goes to the background. Fuzzy Associative Memories (FAM) by Kosko [29], fuzzy rule extraction using self organizing maps by Pedrycz et al. [46] and the systems capable of learning of fuzzy set parameters by Nomura et al. [44] are some good examples of cooperative neuro-fuzzy systems.

3.2.1 Fuzzy Associative Memories

Kosko interprets a fuzzy rule as an association between antecedent and consequent parts [29]. If a fuzzy set is seen as a point in the unit hypercube and rules are associations, then it is possible to use neural associative memories to store fuzzy rules. A neural associative memory can be represented by its connection matrix. Associative recall is equivalent to multiplying a key factor with this matrix. The weights store the correlations between the features of the key k and the information part i. Due to the restricted capacity of associative memories and because of the combination of multiple connection matrices into a single matrix is not recommended due to severe loss of information, it is necessary to store each fuzzy rule in a single FAM. Rules with n conjunctively combined variables in their antecedents can be represented by n FAMs, where each stores a single rule. The FAMs are completed by aggregating all the individual outputs (maximum operator in the case of Mamdani fuzzy system) and a defuzzification component.

Learning could be incorporated in FAM, as learning the weights associated with FAMs output or to create FAMs completely by learning. A neural network-learning algorithm determines the rule weights for the fuzzy rules. Such factors are often interpreted as the influence of a rule and are multiplied with the rule outputs. Rule weights can be replaced equivalently by modifying the membership functions. However, this could result in misinterpretation of fuzzy sets and identical linguistic values might be represented differently in different rules. Kosko suggests a form of adaptive vector quantization technique to learn the FAMs. This approach is termed as differential competitive learning and is very similar to the learning in self-organizing maps.

Figure 3.1 depicts a cooperative neuro-fuzzy model where the neural network learning mechanism is used to determine the fuzzy rules, parameters of fuzzy sets, rule weights etc. Kosko's adaptive FAM is a cooperative neuro-fuzzy model because it uses a learning technique to determine the rules and

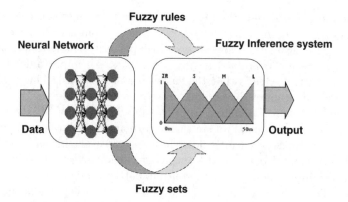

Fig. 3.1. Cooperative neuro-fuzzy model

its weights. The main disadvantage of FAM is the weighting of rules. Just because certain rules, does not have much influence does not mean that they are very unimportant. Hence, the reliability of FAMs for certain applications is questionable. Due to implementation simplicity, FAMs are used in many applications.

3.2.2 Fuzzy Rule Extraction Using Self Organizing Maps

Pedryz et al. [46] used self-organizing maps with a planar competition layer to cluster training data, and they provide means to interpret the learning results. The learning results could show whether two input vectors are similar to each other or belong to the same class. However, in the case of high-dimensional input vectors, the structure of the learning problem can rarely be detected in the two dimensional map. A procedure is provided for interpreting the learning results using linguistic variables.

After the learning process, the weight matrix W represents the weight of each feature of the input patterns to the output. Such a matrix defines a map for a single feature only. For each feature of the input patterns, fuzzy sets are specified by a linguistic description B (one fuzzy set for each variable). They are applied to the weight matrix W to obtain a number of transformed matrices. Each combination of linguistic terms is a possible description of a pattern subset or cluster. To check a linguistic description B for validity, the transformed maps are intersected and a matrix D is obtained. Matrix D determines the compatibility of the learning result with the linguistic description B. $D^{(B)}$ is a fuzzy relation, and $D^{(B)}$ is interpreted as the degree of support of B. By describing $D^{(B)}$ by its α-cuts D_α^B one obtains subsets of output nodes, whose degree of membership is at least α such that the confidence of all patterns X_α belong to the class described by B vanishes with decreasing α. Each B is a valid description of a cluster if $D^{(B)}$ has a non-empty α-cut D_α^B. If the features are separated into input and output features according to the application considered, then each B represents a linguistic rule, and by examining each combination of linguistic values, a complete fuzzy rule base can be created. This method also shows which patterns belong to a fuzzy rule, because they are not contained in any subset X_α. An important advantage when compared to FAMs is that the rules are not weighted. The problem is with the determination of the number of output neurons and the α values for each learning problem. Compared to FAM, since the form of the membership function determines a crucial role in the performance the data could be better exploited. Since Kosko's learning procedure does not take into account of the neighborhood relation between the output neurons, perfect topological mapping from the input patterns to the output patterns might not be obtained sometimes. Thus, the FAM learning procedure is more dependent on the sequence of the training data than Pedryz et al. procedure. The structure of the feature space is initially determined and then the linguistic descriptions best matching the learning results by using the available fuzzy partitions are obtained. If a large number of patterns fit none of the descriptions, this may be due to an insufficient choice of membership functions and they can be determined anew. Hence for learning the fuzzy rules this approach is preferable compared to FAM. Performance of this method still depends on the learning rate and the neighborhood size for weight modification, which is problem dependant and could be determined heuristically. Fuzzy c-means algorithm also has been explored to determine the learning rate and neighborhood size by Bezdek et al. [9].

3.2.3 Systems Capable of Learning Fuzzy Set Parameters

Nomura et al. [44] proposed a supervised learning technique to fine-tune the fuzzy sets of an existing Sugeno type fuzzy system. The learning algorithm uses a gradient descent procedure that uses an error measure E (difference between the actual and target outputs) to fine-tune the parameters of the membership functions (MF). The procedure is very similar to the delta rule for multilayer perceptrons. The learning takes place in an offline mode. For the input vector, the resulting error E is calculated and based on that the consequent parts (a real value) are updated. Then the same patterns are propagated again and only the parameters of the MFs are updated. This is done to take the changes in the consequents into account when the antecedents are modified. A severe drawback of this approach is that the representation of the linguistic values of the input variables depends on the rules they appear in. Initially identical linguistic terms are represented by identical membership functions. During the learning process, they may be developed differently, so that identical linguistic terms are represented by different fuzzy sets. The proposed approach is applicable only to Sugeno type fuzzy inference system. Using a similar approach, Miyoshi et al. [38] adapted fuzzy T-norm and T-conorm operators while Yager et al. [53] adapted the defuzzification operator using a supervised learning algorithm.

3.3 Concurrent Neuro-Fuzzy System

In a concurrent model, neural network assists the fuzzy system continuously (or vice versa) to determine the required parameters especially if the input variables of the controller cannot be measured directly. Such combinations do not optimize the fuzzy system but only aids to improve the performance of the overall system. Learning takes place only in the neural network and the fuzzy system remains unchanged during this phase. In some cases the fuzzy outputs might not be directly applicable to the process. In that case neural network can act as a postprocessor of fuzzy outputs. Figure 3.2 depicts a concurrent neuro-fuzzy model where in the input data is fed to a neural network and the output of the neural network is further processed by the fuzzy system.

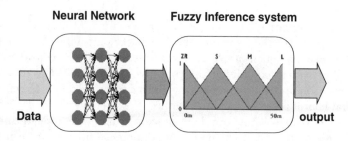

Fig. 3.2. Concurrent neuro-fuzzy model

3.4 Integrated Neuro-Fuzzy Systems

In an integrated model, neural network learning algorithms are used to determine the parameters of fuzzy inference systems. Integrated neuro-fuzzy systems share data structures and knowledge representations. A fuzzy inference system can utilize human expertise by storing its essential components in rule base and database, and perform fuzzy reasoning to infer the overall output value. The derivation of *if-then* rules and corresponding membership functions depends heavily on the a priori knowledge about the system under consideration. However there is no systematic way to transform experiences of knowledge of human experts to the knowledge base of a fuzzy inference system. There is also a need for adaptability or some learning algorithms to produce outputs within the required error rate. On the other hand, neural network learning mechanism does not rely on human expertise. Due to the homogenous structure of neural network, it is hard to extract structured knowledge from either the weights or the configuration of the network. The weights of the neural network represent the coefficients of the hyper-plane that partition the input space into two regions with different output values. If we can visualize this hyper-plane structure from the training data then the subsequent learning procedures in a neural network can be reduced. However, in reality, the a priori knowledge is usually obtained from human experts, it is most appropriate to express the knowledge as a set of fuzzy if-then rules, and it is very difficult to encode into a neural network.

Table 3.1. Comparison between neural networks and fuzzy inference systems

Artificial Neural Network	Fuzzy Inference System
Difficult to use prior rule knowledge	Prior rule-base can be incorporated
Learning from scratch	Cannot learn (linguistic knowledge)
Black box	Interpretable (if-then rules)
Complicated learning algorithms	Simple interpretation and implementation
Difficult to extract knowledge	Knowledge must be available

Table 3.1 summarizes the comparison between neural networks and fuzzy inference system. To a large extent, the drawbacks pertaining to these two approaches seem complementary. Therefore, it seems natural to consider building an integrated system combining the concepts of FIS and ANN modeling. A common way to apply a learning algorithm to a fuzzy system is to represent it in a special neural network like architecture. However the conventional neural network learning algorithms (gradient descent) cannot be applied directly to such a system as the functions used in the inference process are usually non differentiable. This problem can be tackled by using differentiable functions in the inference system or by not using the standard neural learning algorithm. In Sects. 3.4.1 and 3.4.2, we will discuss how to model integrated neuro-fuzzy systems implementing Mamdani [36] and Takagi-Sugeno FIS [47].

3.4.1 Mamdani Integrated Neuro-Fuzzy Systems

A Mamdani neuro-fuzzy system uses a supervised learning technique (backpropagation learning) to learn the parameters of the membership functions. Architecture of Mamdani neuro-fuzzy system is illustrated in Fig. 3.3. The detailed function of each layer is as follows:

Layer-1 *(input layer):* No computation is done in this layer. Each node in this layer, which corresponds to one input variable, only transmits input values to the next layer directly. The link weight in layer 1 is unity.

Layer-2 *(fuzzification layer):* Each node in this layer corresponds to one linguistic label (excellent, good, etc.) to one of the input variables in layer 1. In other words, the output link represent the membership value, which specifies the degree to which an input value belongs to a fuzzy set, is calculated in layer 2. A clustering algorithm will decide the initial number and type of membership functions to be allocated to each of the input variable. The final shapes of the MFs will be fine tuned during network learning.

Layer-3 *(rule antecedent layer):* A node in this layer represents the antecedent part of a rule. Usually a T-norm operator is used in this node. The

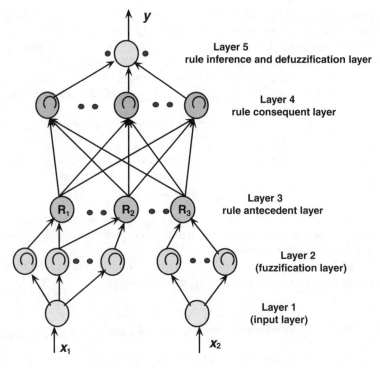

Fig. 3.3. Mamdani neuro-fuzzy system

output of a layer 3 node represents the firing strength of the corresponding fuzzy rule.

Layer-4 *(rule consequent layer):* This node basically has two tasks. To combine the incoming rule antecedents and determine the degree to which they belong to the output linguistic label (high, medium, low, etc.). The number of nodes in this layer will be equal to the number of rules.

Layer-5 *(combination and defuzzification layer):* This node does the combination of all the rules consequents using a T-conorm operator and finally computes the crisp output after defuzzification.

3.4.2 Takagi-Sugeno Integrated Neuro-Fuzzy System

Takagi-Sugeno neuro-fuzzy systems make use of a mixture of backpropagation to learn the membership functions and least mean square estimation to determine the coefficients of the linear combinations in the rule's conclusions. A step in the learning procedure got two parts: In the first part the input patterns are propagated, and the optimal conclusion parameters are estimated by an iterative least mean square procedure, while the antecedent parameters (membership functions) are assumed to be fixed for the current cycle through the training set. In the second part the patterns are propagated again, and in this epoch, backpropagation is used to modify the antecedent parameters, while the conclusion parameters remain fixed. This procedure is then iterated. The detailed functioning of each layer (as depicted in 3.4) is as follows:

Layers 1, 2 and 3 functions the same way as Mamdani FIS.

Layer 4 *(rule strength normalization):* Every node in this layer calculates the ratio of the i-th rule's firing strength to the sum of all rules firing strength

$$\overline{w_i} = \frac{w_i}{w_1 + w_2}, i = 1, 2..$$ (3.1)

Layer-5 *(rule consequent layer):* Every node i in this layer is with a node function

$$\overline{w_i} f_i = \overline{w_i} (p_i\, x_1 + q_i\, x_2 + r_i)$$ (3.2)

where $\overline{w_i}$ is the output of layer 4, and $\{p_i, q_i, r_i\}$ is the parameter set. A well-established way is to determine the consequent parameters using the least means squares algorithm.

Layer-6 *(rule inference layer)* The single node in this layer computes the overall output as the summation of all incoming signals

$$Overall\ output = \sum_i \overline{w_i} f_i = \frac{\sum_i w_i f_i}{\sum_i w_i}$$ (3.3)

In the following sections, we briefly discuss the different integrated neuro-fuzzy models that make use of the complementarities of neural networks and fuzzy inference systems implementing a Mamdani or Takagi Sugeno fuzzy inference system. Some of the major works in this area are GARIC [8], FALCON [32], ANFIS [24], NEFCON [40], NEFCLASS [41], NEFPROX [43], FUN [48], SONFIN [16], FINEST [50], EFuNN [26], dmEFuNN [27], EvoNF [4], and many others [25, 39, 54].

3.4.3 Adaptive Network Based Fuzzy Inference System (ANFIS)

ANFIS is perhaps the first integrated hybrid neuro-fuzzy model [24] and the architecture is very similar to Fig. 3.4. A modified version of ANFIS as shown in Fig. 3.5 is capable of implementing the Tsukamoto fuzzy inference system [24, 51] as depicted in Fig. 3.6. In the Tsukamoto FIS, the overall output is the weighted average of each rule's crisp output induced by the rule's firing strength (the product or minimum of the degrees of match with the premise part) and output membership functions. The output membership functions used in this scheme must be monotonically non-decreasing. The first hidden layer is for fuzzification of the input variables and T-norm operators are deployed in the second hidden layer to compute the rule antecedent part. The third hidden layer normalizes the rule strengths followed by the fourth hidden layer where the consequent parameters of the rule are determined. Output layer computes the overall input as the summation of all incoming signals.

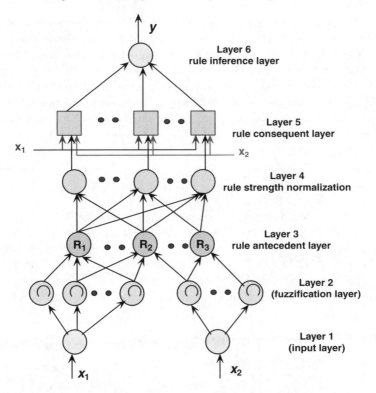

Fig. 3.4. Takagi Sugeno neuro-fuzzy system

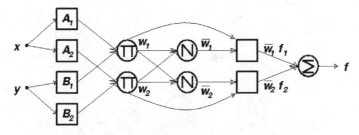

Fig. 3.5. Architecture of ANFIS implementing Tsukamoto fuzzy inference system

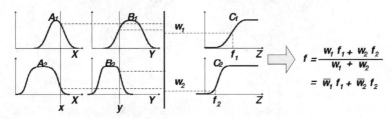

Fig. 3.6. Tsukamoto fuzzy reasoning

In ANFIS, the adaptation (learning) process is only concerned with parameter level adaptation within fixed structures. For large-scale problems, it will be too complicated to determine the optimal premise-consequent structures, rule numbers etc. The structure of ANFIS ensures that each linguistic term is represented by only one fuzzy set. However, the learning procedure of ANFIS does not provide the means to apply constraints that restrict the kind of modifications applied to the membership functions. When using Gaussian membership functions, operationally ANFIS can be compared with a radial basis function network.

3.4.4 Fuzzy Adaptive Learning Control Network (FALCON)

FALCON [32] has a five-layered architecture as shown in Fig. 3.7 and implements a Mamdani type FIS. There are two linguistic nodes for each output variable. One is for training data (desired output) and the other is for the actual output of FALCON. The first hidden layer is responsible for the fuzzification of each input variable. Each node can be a single node representing a simple membership function (MF) or composed of multilayer nodes that

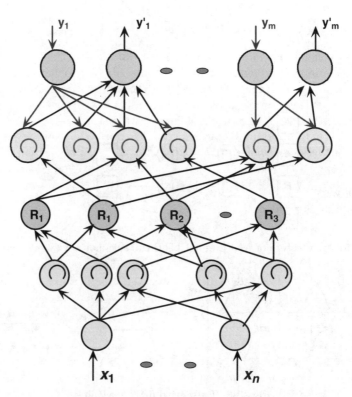

Fig. 3.7. Architecture of FALCON

compute a complex MF. The Second hidden layer defines the preconditions of the rule followed by rule consequents in the third hidden layer. FALCON uses a hybrid-learning algorithm comprising of unsupervised learning and a gradient descent learning to optimally adjust the parameters to produce the desired outputs. The hybrid learning occurs in two different phases. In the initial phase, the centers and width of the membership functions are determined by self-organized learning techniques analogous to statistical clustering techniques. Once the initial parameters are determined, it is easy to formulate the rule antecedents. A competitive learning algorithm is used to determine the correct rule consequent links of each rule node. After the fuzzy rule base is established, the whole network structure is established. The network then enters the second learning phase to adjust the parameters of the (input and output) membership functions optimally. The backpropagation algorithm is used for the supervised learning. Hence FALCON algorithm provides a framework for structure and parameter adaptation for designing neuro-fuzzy systems [32].

3.4.5 Generalized Approximate Reasoning
Based Intelligent Control (GARIC)

GARIC [8] is an extended version of Berenji's Approximate Reasoning based Intelligent Control (ARIC) that implements a fuzzy controller by using several specialized feedforward neural networks [7]. Like ARIC, it consists of an Action state Evaluation Network (AEN) and an Action Selection Network (ASN). The AEN is an adaptive critic that evaluates the actions of the ASN. The ASN does not use any weighted connections, but the learning process modifies parameters stored within the units of the network. Architecture of the GARIC-ASN is depicted in Fig. 3.8. ASN of GARIC is feedforward network with

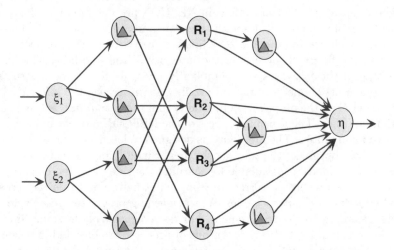

Fig. 3.8. ASN of GARIC

five layers. The first hidden layer stores the linguistic values of all the input variables. Each input unit is only connected to those units of the first hidden layer, which represent its associated linguistic values. The second hidden layer represents the fuzzy rules nodes, which determine the degree of fulfillment of a rule using a *softmin* operation. The third hidden layer represents the linguistic values of the control output variable η. Conclusions of the rule are computed depending on the strength of the rule antecedents computed by the rule node layer. GARIC makes use of local mean-of-maximum method for computing the rule outputs. This method needs a crisp output value from each rule. Therefore, the conclusions must be defuzzified before they are accumulated to the final output value of the controller. The learning algorithm of the AEN of GARIC is equivalent to that of its predecessor ARIC. However, the ASN learning procedure is different from the procedure used in ARIC. GARIC uses a mixture of gradient descent and reinforcement learning to fine-tune the node parameters. The hybrid learning stops if the output of the AEN ceases to change. The interpretation of GARIC is improved compared to GARIC. The relatively complex learning procedure and the architecture of GARIC can be seen as a main disadvantage of GARIC.

3.4.6 Neuro-Fuzzy Controller (NEFCON)

The learning algorithm defined for NEFCON is able to learn fuzzy sets as well as fuzzy rules implementing a Mamdani type FIS [40]. This method can be considered as an extension to GARIC that also use reinforcement learning but need a previously defined rule base. Figure 3.9 illustrates the basic NEFCON architecture with 2 inputs and five fuzzy rules [40]. The inner nodes $R_1, \ldots, R_5$ represent the rules, the nodes ξ_1, ξ_2, and η the input and output values, and μ_r, V_r the fuzzy sets describing the antecedents and consequents. In contrast to neural networks, the connections in NEFCON are weighted with fuzzy sets instead of real numbers. Rules with the same antecedent use so-called shared weights, which are represented by ellipses drawn around the connections as shown in the figure. They ensure the integrity of the rule base. The knowledge base of the fuzzy system is implicitly given by the network structure. The input units assume the task of fuzzification interface, the inference logic is represented by the propagation functions, and the output unit is the defuzzification interface. The learning process of the NEFCON model can be divided into two main phases. The first phase is designed to learn the rule base and the second phase optimizes the rules by shifting or modifying the fuzzy sets of the rules. Two methods are available for learning the rule base. Incremental rule learning is used when the correct out put is not known and rules are created based on estimated output values. As the learning progresses, more rules are added according to the requirement. For decremental rule learning, initially rules are created due to fuzzy partitions of process variables and unnecessary rules are eliminated in the course of learning. Decremental rule learning is less efficient compared to incremental approach. However it can be applied to

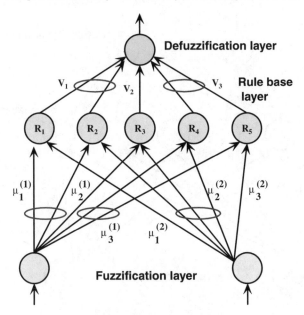

Fig. 3.9. Architecture of NEFCON

unknown processes without difficulty, and there is no need to know or to guess an optimal output value. Both phases use a fuzzy error E, which describes the quality of the current system state, to learn or to optimize the rule base. To obtain a good rule base it must be ensured that the state space of the process is sufficiently covered during the learning process. Due to the complexity of the calculations required, the decremental learning rule can only be used, if there are only a few input variables with not too many fuzzy sets. For larger systems, the incremental learning rule will be optimal. Prior knowledge whenever available could be incorporated to reduce the complexity of the learning. Membership functions of the rule base are modified according to the Fuzzy Error Backpropagation (FEBP) algorithm. The FEBP algorithm can adapt the membership functions, and can be applied only if there is already a rule base of fuzzy rules. The idea of the learning algorithm is identical: increase the influence of a rule if its action goes in the right direction (rewarding), and decrease its influence if a rule behaves counter productively (punishing). If there is absolutely no knowledge about initial membership function, a uniform fuzzy partition of the variables should be used.

3.4.7 Neuro-Fuzzy Classification (NEFCLASS)

NEFCLASS is used to derive fuzzy rules from a set of data that can be separated in different crisp classes [41]. The rule base of a NEFCLASS system approximates an unknown function ϕ that represents the classification problem and maps an input pattern x to its class C_i:

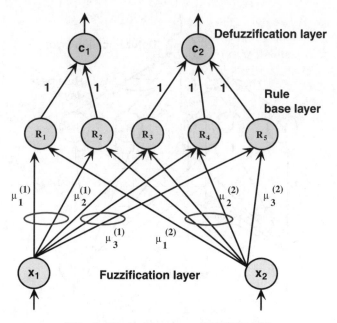

Fig. 3.10. Architecture of NEFCLASS

Because of the propagation procedures used in NEFCLASS the rule base actually does not approximate ϕ but a function ϕ'. We obtain $\phi(x)$ from the equality $\phi(x) = \phi(\phi(x))$, where ϕ reflects the interpretation of the classification result obtained from a NEFCLASS system [42]. Figure 3.10 illustrates the NEFCLASS system that maps patterns with two features into two distinct classes by using five linguistic rules. The NEFCLASS very much resemble the NEFCON system except the slight variation in the learning algorithm and the interpretation of the rules. As in NEFCON system, in NEFCLASS identical linguistic values of an input variable are represented by the same fuzzy set. As classification is the primary task of NEFCLASS, there should be two rules with identical antecedents and each rule unit must be connected to only one output unit. The weights between rule layer and the output layer only connect the units. A NEFCLASS system can be built from partial knowledge about the patterns, and can then be refined by learning, or it can be created from scratch by learning. A user must define a number of initial fuzzy sets that partition the domains of the input features, and specify a value for k, i.e. the maximum number of rule nodes that may be created in the hidden layer. NEFCLASS makes use of triangular membership functions and the learning algorithm of the membership functions uses an error measure that tells whether the degree of fulfillment of a rule has to be higher or lower. This information is used to change the input fuzzy sets. Being a classification system, we are not much interested in the exact output values. In addition, we take a winner-takes-all interpretation for the output, and we are mainly interested in the correct classification result. The incremental rule learning in NEFCLASS is much less expensive than decremental rule learning in NE-FCON. It is possible to build up a rule base in a single sweep through the training set. Even for higher dimensional problems, the rule base is completed after at most three cycles. Compared to neural networks, NEFCLASS uses a much simpler learning strategy. There is no vector quantization involved in finding the rules (clusters, and there is no gradient information needed to train the membership functions. Some other advantages are interpretability, possibility of initialization (incorporating prior knowledge) and its simplicity.

3.4.8 Neuro-Fuzzy Function Approximation (NEFPROX)

NEFPROX system is based on plain supervised learning (fixed learning problem) and it is used for function approximation [43]. It is a modified version of the NEFCON model without the reinforcement learning. NEFPROX (Fig. 3.11) is very much similar to NEFCON and NEFCLASS except the fact that NEFCON have only a single output node and NEFCLASS systems do not use membership functions on the conclusion side. We can initialize the NEFPROX system if we already know suitable rules or else the system is capable to incrementally learn all rules. NEFPROX architecture is as shown in Fig. 11. While ANFIS is capable to implement only Sugeno models with differentiable functions, NEFPROX can learn common Mamdani type of fuzzy

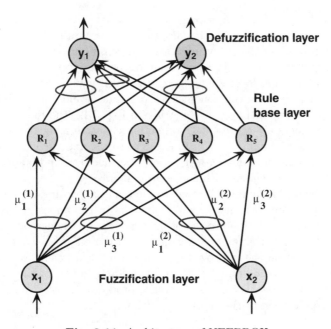

Fig. 3.11. Architecture of NEFPROX

system from data. Further NEFPROX is much faster compared to ANFIS to yield results.

3.4.9 Fuzzy Inference Environment Software with Tuning (FINEST)

FINEST is designed to tune the fuzzy inference itself. FINEST is capable of two kinds of tuning process, the tuning of fuzzy predicates, combination functions and the tuning of an implication function [50]. The three important features of the system are:

- The generalized modus ponens is improved in the following four ways (1) aggregation operators that have synergy and cancellation nature (2) a parameterized implication function (3) a combination function, which can reduce fuzziness (4) backward chaining based on generalized modus ponens.
- Aggregation operators with synergy and cancellation nature are defined using some parameters, indicating the strength of the synergic affect, the area influenced by the effect, etc., and the tuning mechanism is designed to tune also these parameters. In the same way, the tuning mechanism can also tune the implication function and combination function.
- The software environment and the algorithms are designed for carrying out forward and backward chaining based on the improved generalized modus ponens and for tuning various parameters of a system.

FINEST make use of a backpropagation algorithm for the fine-tuning of the parameters. Figure 3.12 shows the layered architecture of FINEST and the calculation process of the fuzzy inference. The input values (x_i) are the facts and the output value (y) is the conclusion of the fuzzy inference. Layer 1 is a fuzzification layer and layer 2 aggregates the truth-values of the conditions of Rule i. Layer 3 deduces the conclusion from Rule I and the combination of all the rules is done in Layer 4. Referring to Fig. 3.12, the function and_i, I_i and $comb$ respectively represent the function characterizing the aggregation operator of rule i, the implication function of rule i, and the global combination function. The functions and_i, I_i, $comb$ and membership functions of each fuzzy predicate are defined with some parameters.

Backpropagation method is used to tune the network parameters. It is possible to tune any parameter, which appears in the nodes of the network representing the calculation process of the fuzzy data if the derivative function with respect to the parameters is given. Thus, FINEST framework provides a mechanism based on the improved generalized modus ponens for fine tuning of fuzzy predicates and combination functions and tuning of the implication function. Parameterization of the inference procedure is very much essential for proper application of the tuning algorithm.

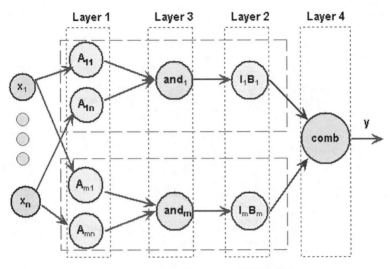

Fig. 3.12. Architecture of FINEST

3.4.10 Self Constructing Neural Fuzzy Inference Network (SONFIN)

SONFIN implements a Takagi-Sugeno type fuzzy inference system. Fuzzy rules are created and adapted as online learning proceeds via a simultaneous structure and parameter identification [16]. In the structure identification of the precondition part, the input space is partitioned in a flexible way according to an aligned clustering based algorithm. As to the structure identification of the consequent part, only a singleton value selected by a clustering method is assigned to each rule initially. Afterwards, some additional significant terms (input variables) selected via a projection-based correlation measure for each rule will be added to the consequent part (forming a linear equation of input variables) incrementally as learning proceeds. For parameter identification, the consequent parameters are tuned optimally by either Least Mean Squares [LMS] or Recursive Least Squares [RLS] algorithms and the precondition parameters are tuned by back propagation algorithm. To enhance knowledge representation ability of SONFIN, a linear transformation for each input variable can be incorporated into the network so that much fewer rules are needed or higher accuracy can be achieved. Proper linear transformations are also learned dynamically in the parameter identification phase of SONFIN. Figure 3.13 illustrates the 6-layer structure of SONFIN.

Learning progresses concurrently in two stages for the construction of SONFIN. The *structure* learning includes both the precondition and consequent structure identification of a fuzzy *if-then* rule. The parameter learning is based on supervised learning algorithms, the parameters of the linear equations in the consequent parts are adjusted by either LMS or RLS algorithms

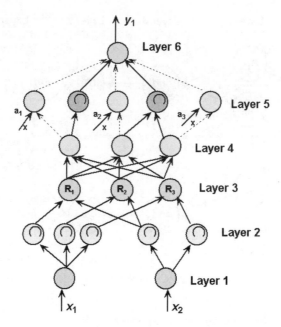

Fig. 3.13. Six layered architecture of SONFIN

and the parameters in the precondition part are adjusted by the backpropagation algorithm. SONFIN can be used for normal operation at anytime during the learning process without repeated training on the input-output pattern when online operation is required. In SONFIN rule base is dynamically created as the learning progresses by performing the following learning processes:

• **Input-output space partitioning**

The way the input space is partitioned determines the number of rules extracted from the training data as well as the number of fuzzy sets on the universal of discourse of each input variable. For each incoming pattern x the strength a rule is fired can be interpreted as the degree the incoming pattern belongs to the corresponding cluster. The center and width of the corresponding membership functions (of the newly formed fuzzy rules) are assigned according to the first-neighbor heuristic. For each rule generated, the next step is to decompose the multidimensional membership function to corresponding $1 - D$ membership function for each input variable. For the output space partitioning, almost a similar measure is adopted. Performance of SONFIN can be enhanced by incorporating a transformation matrix R into the structure, which accommodates all the *a priori* knowledge of the data set.

• **Construction of fuzzy rule base**

Generation of new input cluster corresponds to the generation of a new fuzzy rule, with its precondition part constructed by the learning algorithm in process. At the same time we have to decide the consequent part of the generated rule. This is done using a algorithm based on the fact that different preconditions of rules may be mapped to the same consequent fuzzy set. Since only the center of each output membership function is used for defuzzification, the consequent part of each rule may simply be regarded as a singleton. Compared to the general fuzzy rule based models with singleton output where each rule has its own singleton value, fewer parameters are needed in the consequent part of the SONFIN, especially for complicated systems with a large number of rules.

- **Optimal consequent structure identification**

TSK model can model a sophisticated system with a few rules. In SONFIN, instead of using the linear combination of all input variables as the consequent part, only the most significant input variables are used as the consequent terms of the SONFIN. The significant terms will be chosen and added to the network incrementally any time when the parameter learning cannot improve the network output accuracy anymore during the online learning process. The consequent structure identification scheme in SONFIN is a kind of node growing method in ANNs. When the effect of the parameter learning diminished (output error is not decreasing), additional terms are added to the consequent part.

- **Parameter identification**

After the network structure is adjusted according to the current training pattern, the network then enters the parameter identification phase to adjust the parameters of the network optimally based on the same training pattern. Parameter learning is performed on the whole network after structure learning, no matter whether the nodes (links) are newly added or are existent originally. Backpropagation algorithm is used for this supervised learning. SONFIN is perhaps one of the most computational expensive among all neuro-fuzzy models. The network is adaptable to the users specification of required accuracy.

3.4.11 Fuzzy Net (FUN)

In FUN in order to enable an unequivocal translation of fuzzy rules and membership functions into the network, special neurons have been defined, which, through their activation functions, can evaluate logic expressions [48]. The network consists of an input, an output and three hidden layers. The neurons of each layer have different activation functions representing the different stages in the calculation of fuzzy inference. The activation function can be individually chosen for problems. The network is initialized with a fuzzy rule base and the corresponding membership functions. Figure 14 illustrates the FUN network. The input variables are stored in the input neurons. The neurons in the first hidden layer contain the membership functions and this performs a

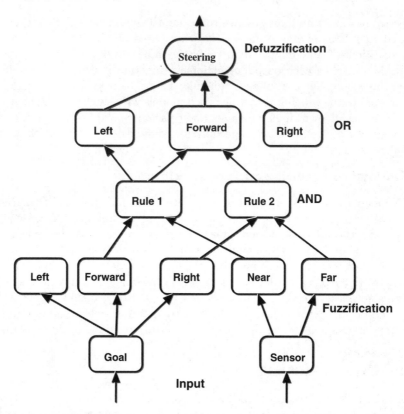

Fig. 3.14. Architecture of the FUN showing the implementation of a sample rule

fuzzification of the input values. In the second hidden layer, the conjunctions (fuzzy-AND) are calculated. Membership functions of the output variables are stored in the third hidden layer. Their activation function is a fuzzy-OR. Finally, the output neurons contain the output variables and have a defuzzification activation function. FUN network is depicted in Fig. 3.14.

Rule: *IF (Goal IS forward AND Sensor IS near) OR (goal IS right AND sensor IS far) THEN steering = forward*

The rules and the membership functions are used to construct an initial FUN network. The rule base can then be optimized by changing the structure of the net or the data in the neurons. To learn the rules, the connections between the rules and the fuzzy values are changed. To learn the membership functions, the data of the nodes in the first and three hidden layers are changed. FUN can be trained with the standard neural network training strategies such as reinforcement or supervised learning.

- **Learning of the rules and membership functions**

The rules are represented in the net through the connections between the layers. The learning of the rules is implemented as a stochastic search in the rule space: a randomly chosen connection is changed and the new network performance is verified with a cost function. If the performance is worse, the change is undone, otherwise it is kept and some other changes are tested, until the desired output is achieved. As the learning algorithm should preserve the semantic of the rules, it has to be controlled in such a way that no two values of the same variable appear in the same rule. This is achieved by swapping connections between the values of the same variable. FUN uses a mixture of gradient descent and stochastic search for updating the membership functions. A maximum change in a random direction is initially assigned to all Membership function Descriptors (MFDs). In a random fashion one MFD of one linguistic variable is selected, and the network performance is tested with this MFD altered according to the allowable change for this MFD. If the network performs better according to the given cost function, the new value is accepted and next time another change is tried in the same direction. Contrary if the network performs worse, the change is reversed. To guarantee convergence, the changes are reduced after each training step and shrink asymptotically towards zero according to the learning rate. As evident, FUN system is initialized by specifying a fixed number of rules and a fixed number of initial fuzzy sets for each variable and the network learns through a stochastic procedure that randomly changes parameters of membership functions and connections within the network structure Since no formal neural network learning technique is used it is questionable to call FUN a neuro-fuzzy system.

3.4.12 Evolving Fuzzy Neural Networks (EFuNN)

EFuNNs [26] and dmEFuNNs [27] are based on the ECOS (Evolving COnnectionist Systems) framework for adaptive intelligent systems formed because of evolution and incremental, hybrid (supervised/unsupervised), online learning. They can accommodate new input data, including new features, new classes, and etc. through local element tuning.

In EFuNNs all nodes are created during learning. EFuNN has a five-layer architecture as shown in Fig. 3.15. The input layer is a buffer layer representing the input variables. The second layer of nodes represents fuzzy quantification of each input variable space. Each input variable is represented here by a group of spatially arranged neurons to represent a fuzzy quantization of this variable. The nodes representing membership functions (triangular, Gaussian, etc) can be modified during learning. The third layer contains rule nodes that evolve through hybrid supervised/unsupervised learning. The rule nodes represent prototypes of input-output data associations, graphically represented as an association of hyper-spheres from the fuzzy input and fuzzy output spaces. Each rule node r is defined by two vectors of connection weights: $W_1(r)$ and

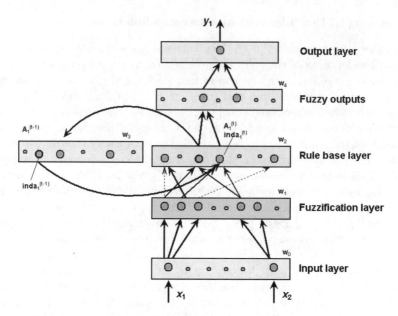

Fig. 3.15. Architecture of EFuNN

$W_2(r)$, the latter being adjusted through supervised learning based on the output error, and the former being adjusted through unsupervised learning based on similarity measure within a local area of the input problem space. The fourth layer of neurons represents fuzzy quantification for the output variables. The fifth layer represents the real values for the output variables. In the case of "one-of-n" EFuNNs, the maximum activation of the rule node is propagated to the next level. In the case of *"many-of-n"* mode, all the activation values of rule nodes that are above an activation threshold are propagated further in the connectionist structure.

3.4.13 Dynamic Evolving Fuzzy Neural Networks (dmEFuNNs)

Dynamic Evolving Fuzzy Neural Networks (dmEFuNN) model is developed with the idea that not just the winning rule node's activation is propagated but a group of rule nodes is dynamically selected for every new input vector and their activation values are used to calculate the dynamical parameters of the output function. While EFuNN make use of the weighted fuzzy rules of Mamdani type, dmEFuNN uses the Takagi-Sugeno fuzzy rules. The architecture is depicted in Fig. 3.16.

The first, second and third layers of dmEFuNN have exactly the same structures and functions as the EFuNN. The fourth layer, the fuzzy inference layer, selects m rule nodes from the third layer which have the closest fuzzy normalised local distance to the fuzzy input vector, and then, a TakagiSugeno

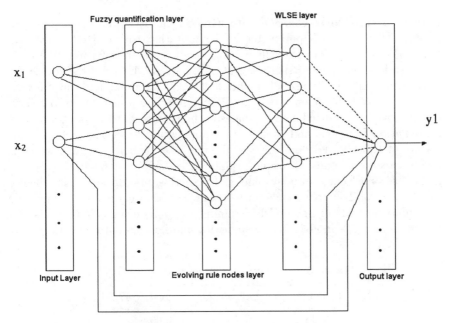

Fig. 3.16. Architecture of dmEFuNN

fuzzy rule will be formed using the weighted least square estimator. The last layer calculates the output of dmEFuNN.

The number m of activated nodes used to calculate the output values for a dmEFuNN is not less than the number of the input nodes plus one. Like the EFuNNs, the dmEFuNNs can be used for both offline learning and online learning thus optimising global generalization error, or a local generalization error. In dmEFuNNs, for a new input vector (for which the output vector is not known), a subspace consisted of m rule nodes are found and a first order TakagiSugeno fuzzy rule is formed using the least square estimator method. This rule is used to calculate the dmEFuNN output value. In this way, a dmE-FuNN acts as a universal function approximator using m linear functions in a small m dimensional node subspace. The accuracy of approximation depends on the size of the node subspaces, the smaller the subspace is, the higher the accuracy. It means that if there are sufficient training data vectors and sufficient rule nodes are created, a satisfying accuracy can be obtained.

3.5 Discussions

As evident, both cooperative and concurrent models are not fully interpretable due to the presence of neural network (black box concept). Whereas an integrated neuro-fuzzy model is interpretable and capable of learning in a supervised mode (or even reinforcement learning like NEFCON). In FALCON,

GARIC, ANFIS, NEFCON, SONFIN, FINEST and FUN the learning process is only concerned with parameter level adaptation within fixed structures. For large-scale problems, it will be too complicated to determine the optimal premise-consequent structures, rule numbers etc. User has to provide the architecture details (type and quantity of MF's for input and output variables), type of fuzzy operators etc. FINEST provides a mechanism based on the improved generalized modus ponens for fine tuning of fuzzy predicates and combination functions and tuning of an implication function. An important feature of EFuNN and dmEFuNN is the one pass (epoch) training, which is highly capable of online learning. Table 3.2 provides a comparative performance of some neuro fuzzy systems for predicting the Mackey-Glass chaotic time series [35]. Due to unavailability of source codes, we are unable to provide a comparison with all the models. Training was done using 500 data sets and the considered NF models were tested with another 500 data sets [1].

Table 3.2. Performance of neuro-fuzzy systems

System	Epochs	Test RMSE
ANFIS	75	0.0017
NEFPROX	216	0.0332
EFuNN	1	0.0140
dmEFuNN	1	0.0042
SONFIN	1	0.0180

Among NF models ANFIS has the lowest Root Mean Square Error (RMSE) and NEPROX the highest. This is probably due to Takagi-Sugeno rules implementation in ANFIS compared to the Mamdani-type fuzzy system in NEFPROX. However, NEFPROX outperformed ANFIS in terms of computational time. Due to fewer numbers of rules SONFIN, EFuNN and dmEFuNN are also able to perform faster than ANFIS. Hence, there is a tradeoff between interpretability and accuracy. Takagi Sugeno type inference systems are more accurate but require more computational effort. While Mamdani type inference, systems are more interpretable and required less computational load but often with a compromise on accuracy.

As the problem become, more complicated manual definition of NF architecture/parameters becomes complicated. The following questions remain unanswered:

- For input/output variables, what are the optimal number of membership functions and shape?
- What is the optimal structure (rule base) and fuzzy operators?
- What are the optimal learning parameters?
- Which fuzzy inference system (example. Takagi-Sugeno, Mamdani etc.) will work the best for a given problem?

3.5.1 Evolutionary and Neural Learning of Fuzzy Inference System (EvoNF)

In an integrated neuro-fuzzy model there is no guarantee that the neural network learning algorithm converges and the tuning of fuzzy inference system will be successful. Natural intelligence is a product of evolution. Therefore, by mimicking biological evolution, we could also simulate high-level intelligence. Evolutionary computation works by simulating a population of individuals, evaluating their performance, and evolving the population a number of times until the required solution is obtained. The drawbacks pertaining to neural networks and fuzzy inference systems seem complementary and evolutionary computation could be used to optimize the integration to produce the best possible synergetic behavior to form a single system. Adaptation of fuzzy inference systems using evolutionary computation techniques has been widely explored [14]. EvoNF [4] is an adaptive framework based on evolutionary computation and neural learning wherein the membership functions, rule base and fuzzy operators are adapted according to the problem. The evolutionary search of MFs, rule base, fuzzy operators etc. would progress on different time scales to adapt the fuzzy inference system according to the problem environment. Membership functions and fuzzy operators would be further fine-tuned using a neural learning technique. Optimal neural learning parameters will be decided during the evolutionary search process. Figure 3.17 illustrates the general interaction mechanism of the EvoNF framework with the evolutionary search of fuzzy inference system (Mamdani, Takagi -Sugeno etc.) evolving at the highest level on the slowest time scale. For each evolutionary search of fuzzy operators (best combination of T-norm and T-conorm, defuzzification strategy etc), the search for the fuzzy rule base progresses at a faster time scale in an environment decided by the problem. In a similar manner, evolutionary search of membership functions proceeds at a faster time scale (for every rule base) in the environment decided by the problem. Hierarchy of the different adaptation procedures will rely on the prior knowledge. For example, if there

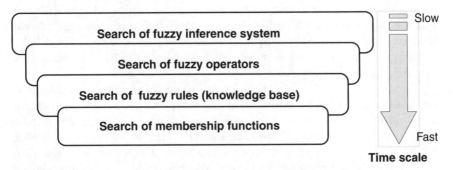

Fig. 3.17. Interaction of evolutionary search mechanisms in the adaptation of fuzzy inference system

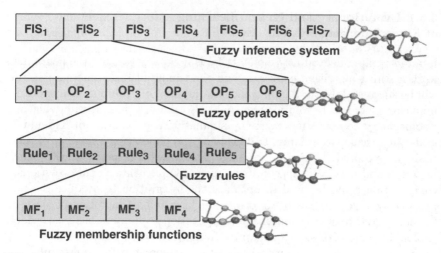

Fig. 3.18. Chromosome representation of the adaptive fuzzy inference system using evolutionary computation and neural learning

is more prior knowledge about the fuzzy rule base than the fuzzy operators then it is better to implement the search for fuzzy rule base at a higher level. The problem representation (genetic coding) is illustrated in Fig. 3.18. Please refer [4] for more technical details.

Automatic adaptation of membership functions is popularly known as self tuning and the genome encodes parameters of trapezoidal, triangle, logistic, hyperbolic-tangent, Gaussian membership functions etc.

Evolutionary search of fuzzy rules can be carried out using three approaches. In the first method, (Michigan approach) the fuzzy knowledge base is adapted because of antagonistic roles of competition and cooperation of fuzzy rules [14]. Each genotype represents a single fuzzy rule and the entire population represents a solution. A classifier rule triggers whenever its condition part matches the current input, in which case the proposed action is sent to the process to be controlled. The global search algorithm will generate new classifier rules based on the rule strengths acquired during the entire process. The fuzzy behavior is created by an activation sequence of mutually collaborating fuzzy rules. The entire knowledge base is build up by a cooperation of competing multiple fuzzy rules.

The second method (Pittsburgh approach) evolves a population of knowledge bases rather than individual fuzzy rules [14]. Genetic operators serve to provide a new combination of rules and new rules. In some cases, variable length rule bases are used; employing modified genetic operators for dealing with these variable length and position independent genomes. The disadvantage is the increased complexity of search space and additional computational burden especially for online learning.

The third method (iterative rule learning approach) is very much similar to the first method with each chromosome representing a single rule, but contrary to the Michigan approach, only the best individual is considered to form part of the solution, discarding the remaining chromosomes in the population. The evolutionary learning process builds up the complete rule base through a iterative learning process [18].

3.6 Conclusions

In this chapter, we presented the different ways to learn fuzzy inference systems using neural network learning techniques. As a guideline, for neuro-fuzzy systems to be highly intelligent some of the major requirements are fast learning (memory based - efficient storage and retrieval capacities), on-line adaptability (accommodating new features like inputs, outputs, nodes, connections etc), achieve a global error rate and computationally inexpensive. The data acquisition and preprocessing training data is also quite important for the success of neuro-fuzzy systems. Many neuro-fuzzy models use supervised/unsupervised techniques to learn the different parameters of the inference system. The success of the learning process is not guaranteed, as the designed model might not be optimal. Empirical research has shown that gradient descent technique (most commonly used supervised learning algorithm) is trapped in local optima especially when the error surface is complicated.

Global optimization procedures like evolutionary algorithms, simulated annealing, tabu search etc. might be useful for adaptive evolution of fuzzy if-then rules, shape and quantity of membership functions, fuzzy operators and other node functions, to prevent the network parameters being trapped in local optima due to reliance on gradient information by most of the supervised learning techniques. For online learning, global optimization procedures might sound computational expensive. Fortunately, evolutionary algorithms work with a population of independent solutions, which makes it easy to distribute the computational load among several processors using parallel algorithms.

Sugeno-type fuzzy systems are high performers (less RMSE) but often requires complicated learning procedures and computational expensive. However, Mamdani-type fuzzy systems can be modeled using faster heuristics but with a compromise on the performance (accuracy). Hence there is always a compromise between performance and computational time.

3.7 Acknowledgements

The author wishes to thank the anonymous reviewers for their constructive comments which helped to improve the presentation of the chapter.

References

1. A. Abraham, Neuro-Fuzzy Systems: State-of-the-Art Modeling Techniques, Connectionist Models of Neurons, Learning Processes, and Artificial Intelligence, Springer-Verlag Germany, Jose Mira and Alberto Prieto (Eds.), Granada, Spain, pp. 269–276, 2001.
2. A. Abraham, Intelligent Systems: Architectures and Perspectives, Recent Advances in Intelligent Paradigms and Applications, Abraham A., Jain L. and Kacprzyk J. (Eds.), Studies in Fuzziness and Soft Computing, Springer Verlag Germany, Chap. 1, pp. 1–35, 2002.
3. A. Abraham and M.R. Khan, Neuro-Fuzzy Paradigms for Intelligent Energy Management, Innovations in Intelligent Systems: Design, Management and Applications, Abraham A., Jain L. and Jan van der Zwaag B. (Eds.), Studies in Fuzziness and Soft Computing, Springer Verlag Germany, Chap. 12, pp. 285–314, 2003.
4. A. Abraham, EvoNF: A Framework for Optimization of Fuzzy Inference Systems Using Neural Network Learning and Evolutionary Computation, The 17th IEEE International Symposium on Intelligent Control, ISIC'02, IEEE Press, pp. 327–332, 2002.
5. P. Andlinger and E.R. Reichl, Fuzzy-Neunet: A Non Standard Neural Network, In Prieto et al., pp. 173–180, 1991.
6. M. Arao, T. Fukuda and K. Shimokima, Flexible Intelligent System based on Fuzzy Neural Networks and Reinforcement Learning, In proceedings of IEEE International Conference on Fuzzy Systems, Vol 5(1), pp. 69–70, 1995.
7. H.R. Berenji and P. Khedkar, Fuzzy Rules for Guiding Reinforcement Learning, In International. Conference on Information Processing and Management of Uncertainty in Knowledge-Based Systems (IPMU'92), pp. 511–514, 1992.
8. H.R. Berenji and P. Khedkar, Learning and Tuning Fuzzy Logic Controllers through Reinforcements, IEEE Transactions on Neural Networks, Vol (3), pp. 724–740, 1992.
9. J.C. Bezdek and S.K. Pal, Fuzzy Models for Pattern Recognition, IEEE Press, New York, 1992.
10. M. Brown, K. Bossley and D. Mills, High Dimensional Neurofuzzy Systems: Overcoming the Course of Dimensionality, In Proceedings. IEEE International. Conference on Fuzzy Systems, pp. 2139–2146, 1995.
11. J.J. Buckley and Y. Hayashi, Hybrid neural nets can be fuzzy controllers and fuzzy expert systems, Fuzzy Sets and Systems, 60: pp. 135–142, 1993.
12. H. Bunke and A. Kandel, Neuro-Fuzzy Pattern Recognition, World Scientific Publishing CO, Singapore, 2000.
13. G.A. Carpenter, S. Grossberg, N. Markuzon, J.H. Reynolds, and D.B. Rosen, Fuzzy ARTMAP: A Neural Network Architecture for Incremental Supervised Learning of Analog Multidimensional Maps, IEEE Transactions Neural Networks, 3(5), pp. 698–712, 1992.
14. O. Cordón F. Herrera, F. Hoffmann and L. Magdalena, Genetic Fuzzy Systems: Evolutionary Tuning and Learning of Fuzzy Knowledge Bases, World Scientific Publishing Company, Singapore, 2001.
15. F. De Souza, M.M.R. Vellasco, M.A.C. Pacheco, The Hierarchical Neuro-Fuzzy BSP Model: An Application in Electric Load Forecasting, Connectionist Models of Neurons, Learning Processes and Artificial Intelligence, Jose Mira et al (Editors), LNCS 2084, Springer Verlag Germany, 2001.

16. J.C. Feng and L.C. Teng, An Online Self Constructing Neural Fuzzy Inference Network and its Applications, IEEE Transactions on Fuzzy Systems, Vol 6, No.1, pp. 12–32, 1998.
17. R. Fuller, Introduction to Neuro-Fuzzy Systems, Studies in Fuzziness and Soft Computing, Springer Verlag, Germany, 2000.
18. A. Gonzalez and F. Herrera, Multi-Stage Genetic Fuzzy Systems Based on the Iterative Rule Learning Approach, Mathware and Soft Computing Vol 4, pp. 233–249, 1997.
19. M.M. Gupta, Fuzzy Neural Computing Systems, In Proceedings of SPIE, Vol 1710, Science of Artificial Neural Networks, Vol 2, pp. 489–499, 1992.
20. S.K. Halgamuge and M. Glesner, Neural Networks in Designing Fuzzy Systems for Real World Applications, Fuzzy Sets and Systems, 65: pp. 1–12, 1994.
21. Y. Hayashi and J.J. Buckley, Approximations Between Fuzzy Expert Systems and Neural Networks, International Journal of Approximate Reasoning, Vol 10, pp. 63–73, 1994.
22. Y. Hayashi and A. Imura, Fuzzy Neural Expert System with Automated Extraction of Fuzzy If-Then Rules from a Trained Neural Network, In First International. Symposium on Uncertainty Modeling and Analysis, pp. 489–494, 1990.
23. J. Hollatz, Neuro-Fuzzy in Legal Reasoning, In Proceedings. IEEE International. Conference on Fuzzy Systems, pp. 655–662, 1995.
24. R. Jang, Neuro-Fuzzy Modeling: Architectures, Analyses and Applications, Ph.D. Thesis, University of California, Berkeley, 1992.
25. A. Kandel, Q.Y. Zhang and H. Bunke, A Genetic Fuzzy Neural Network for Pattern Recognition, In IEEE Transactions on Fuzzy Systems, pp. 75–78, 1997.
26. N. Kasabov, Evolving Fuzzy Neural Networks – Algorithms, Applications and Biological Motivation, In Yamakawa T and Matsumoto G (Eds), Methodologies for the Conception, Design and Application of Soft Computing, World Scientific, pp. 271–274, 1998.
27. N. Kasabov and S. Qun, Dynamic Evolving Fuzzy Neural Networks with m-out-of-n Activation Nodes for On-line Adaptive Systems, Technical Report TR99/04, Department of information science, University of Otago, New Zealand, 1999.
28. E. Khan and P. Venkatapuram, Neufuz: Neural Network Based Fuzzy Logic Design Algorithms, In Proceedings IEEE International Conference on Fuzzy Systems, pp. 647–654, 1993.
29. B. Kosko, Neural Networks and Fuzzy Systems: A Dynamical Systems Approach to Machine Intelligence, Prentice Hall, Englewood Cliffs, New Jersey, 1992.
30. W. Li, Optimization of a Fuzzy Controller Using Neural Network, In Proceedings IEEE International Conference on Fuzzy Systems, pp. 223–227, 1994.
31. X.H. Li and C.L.P. Chen, The Equivalance Between Fuzzy Logic Systems and Feedforward Neural Networks, IEEE Transactions on Neural Networks, Vol 11, No. 2, pp. 356–365, 2000.
32. C.T. Lin and C.S.G. Lee, Neural Network based Fuzzy Logic Control and Decision System, IEEE Transactions on Comput. (40(12): pp. 1320–1336, 1991.
33. C.T. Lin and C.S.G. Lee, Neural Fuzzy Systems: A Neuro-Fuzzy Synergism to Intelligent Systems, Prentice Hall Inc, USA, 1996.

34. A. Lotfi, Learning Fuzzy Inference Systems, Ph.D. Thesis, Department of Electrical and Computer Engineering, University of Queensland, Australia, 1995.
35. M.C. Mackey and L. Glass, Oscillation and Chaos in Physiological Control Systems, Science Vol 197, pp. 287–289, 1977.
36. E.H. Mamdani and S. Assilian, An Experiment in Linguistic Synthesis with a Fuzzy Logic Controller, International Journal of Man-Machine Studies, Vol 7, No.1, pp. 1–13, 1975.
37. S. Mitra and Y. Hayashi, Neuro-Fuzzy Rule Generation: Survey in Soft Computing Framework', IEEE Transactions on Neural Networks, Vol II, No. 3. pp. 748–768, 2000.
38. T. Miyoshi, S. Tano, Y. Kato and T. Arnould, Operator Tuning in Fuzzy Production Rules Using Neural networks, In Proceedings of the IEEE International Conference on Fuzzy Systems, San Francisco, pp. 641–646, 1993.
39. M. Mizumoto and S. Yan, A New Approach of Neurofuzzy Learning Algorithm, Intelligent Hybrid Systems: Fuzzy Logic, Neural Networks, and Genetic Algorithms, Ruan D (Ed.), Kluwer Academic Publishers, pp. 109–129, 1997.
40. D. Nauck and R. Kruse, NEFCON-I: An X-Window Based Simulator for Neural Fuzzy Controllers. In Proceedings of the IEEE International Conference on Neural Networks, Orlando, pp. 1638–1643, 1994.
41. D. Nauck and R. Kruse, NEFCLASS: A Neuro-Fuzzy Approach for the Classification of Data, In Proceedings of ACM Symposium on Applied Computing, George K et al (Eds.), Nashville, ACM Press, pp. 461–465, 1995.
42. D. Nauck and R. Kruse, A Neuro-Fuzzy Method to Learn Fuzzy Classification Rules from Data. Fuzzy Sets and Systems, 89, pp. 277–288, 1997.
43. D. Nauck D. and R. Kruse, Neuro-Fuzzy Systems for Function Approximation, Fuzzy Sets and Systems, 101, pp. 261–271, 1999.
44. H. Nomura, I. Hayashi and N. Wakami, A Learning Method of Fuzzy Inference Systems by Descent Method, In Proceedings of the First IEEE International conference on Fuzzy Systems, San Diego, USA, pp. 203–210, 1992.
45. S.K. Pal and S. Mitra, Neuro-Fuzzy Pattern Recognition: Methods in Soft Computing, John Wiley & Sons, Inc, USA, 1999.
46. W. Pedrycz and H.C. Card, Linguistic Interpretation of Self Organizing Maps, In Prroceedings of the IEEE International Conference on Fuzzy Systems, San Diego, pp. 371–378, 1992.
47. M. Sugeno, Industrial Applications of Fuzzy Control, Elsevier Science Pub Co., 1985.
48. S.M. Sulzberger, N.N. Tschicholg-Gurman, S.J. Vestli, FUN: Optimization of Fuzzy Rule Based Systems Using Neural Networks, In Proceedings of IEEE Conference on Neural Networks, San Francisco, pp. 312–316, 1993.
49. H. Takagi, Fusion Technology of Fuzzy Theory and Neural Networks - Survey and Future Directions, In Proceedings 1st International Conference on Fuzzy Logic & Neural Networks, pp. 13–26, 1990.
50. S. Tano, T. Oyama and T. Arnould, Deep combination of Fuzzy Inference and Neural Network in Fuzzy Inference, Fuzzy Sets and Systems, 82(2) pp. 151–160, 1996.
51. Y. Tsukamoto, An Approach to Fuzzy Reasoning Method, Gupta M.M. et al (Eds.), Advances in Fuzzy Set Theory and Applications, pp. 137–149, 1979.
52. R.R. Yager, On the Interface of Fuzzy Sets and Neural Networks, In International Workshop on Fuzzy System Applications, pp. 215–216, 1988.

53. R.R. Yager and D.P. Filev, Adaptive Defuzzification for Fuzzy System Modeling, In Proceedings of the Workshop of the North American Fuzzy Information Processing Society, pp. 135–142, 1992.
54. Q.Y. Zhang and A. Kandel, Compensatory Neuro-fuzzy Systems with Fast Learning Algorithms, IEEE Transactions on Neural Networks, Vol 9, No. 1, pp. 83–105, 1998.

Part II

Fuzzy Systems

4

A Fuzzy Approach on Guiding Model for Interception Flight

S. Ionita

Department of Electronics and Computer Engineering,
Faculty of Electronics and Electromechanics,
University of Pitesti, 1, Targul din vale, 0300
Pitesti, jud Arges, Romania,
ionis@upit.ro, www.upit.ro

This chapter presents an original moving control model based on the fuzzy logic, applied to some navigation special issues. The steps taken on special flight tasks prove adequacy of the fuzzy rules based model to this field. The problems of guiding on interception trajectories (applied to missiles and air fighting), of getting closer and flight coupling (applied to flight refueling or spacecrafts coupling) and of landing are candidate issues in order to be dealt with from the fuzzy perspective. The steps of this chapter are developed from a comparative study perspective involving one of the most popular techniques such as proportional navigation and the proposed fuzzy based method. Three particular forms of the proportional technique are compared against the fuzzy model. One of these particular proportional methods joins the analytic geometric relationship with the fuzzy controlled speed of the interceptor. The fuzzy fully controlled navigation in the proposed model does not impose any analytic relationship between the positions of the aircrafts. The chapter describes the methodology followed on fuzzy controller developing. The results emphasize performances of the fuzzy control on simulation cases and the advantages of implementing this type of automatic equipment.

4.1 Introduction

For years the field of aeronautical engineering has been one of the significant domains where traditionally human-in-the-loop control coexists with total automatic flight-control solutions [1,8]. On one hand, the personnel involved in this kind of tasks have to rely on proven methods and much experience. One the other hand, there is the need to develop the appropriate control solutions in order to cope with the complexity of the processes. Nowadays, a very common approach of dynamic processes control consists in the use of models specific to artificial intelligence (AI) and its related paradigm computational intelligence (CI). Advanced CI-based approaches in the aeronautical engineering aim to meet on one hand the basic requirements of control: real time response, robustness, optimality and flexibility, and the specific requirements of current flight mission, on the other hand. Models resulting from mathematical formalization of the physical phenomena that take place in flight tasks usually support description of these kinds of dynamics. However, there are many non standard situations where the increased analytical complexity of these models makes them difficult to solve and very limited in applications. Control models based on AI are critical especially in "untailored dynamics". In what we call "unconventional flight" the critical regimes and real disturbances are complex and almost too difficult to be tailored by the traditional analytical methods. The multitude of processes relating to flight is characterized by non-linearity, information uncertainty, random disturbances and of course by the human subjectivity. As a matter of fact, the trend to fully automation flight requires models with a "human consistency". Thus, the flight control systems should have the ability of learning and using previous experience. One way of developing these systems consists in the knowledge based approaches in form of the expert systems (ES) embedded on board of aircrafts and/or in the ground-based flight control centers [1]. Neural networks (NN) are also CI techniques, where the learned knowledge is represented in a set of networked weights, which give valuable control solutions in many flight control application [9,10]. Unlike the above mentioned techniques, the fuzzy knowledge based systems (FKBS) are very promising models for control tasks in aerospace engineering. Despite of the fact that domain of fuzzy systems as theory is still far from his maturation there are few already proven achievements for commercial applications. The fuzzy washing-machine, the fuzzy controlled camera-focusing, the air condition fuzzy controller, the fuzzy refrigerator/heating system and the fuzzy assisted transmission are well known marks of the state of the art in fuzzy logic technology. A brief survey of evolution from fuzzy set theory to computational intelligence is enough to encourage the scientific community and engineers finding the best solutions to largely implement the fuzzy systems in the industrial application and in the large scale distributed systems [11].

In this chapter we propose an original approach to model the kinematics of two aircrafts (the target and the interceptor, respectively) that are involved in an intercepting mission. Basically, this issue is a conventional moving control problem, but it considers (covers) a larger class of dynamic processes. The aim of this chapter has two main issues. First is to attempt developing the new guiding laws for aircraft flight based on fuzzy logic as the better alternative to analytic one. Second, this work provides a comparative study concerning fuzzy guiding methods and few particular analytic methods. For this issue the intensive soft computing techniques on the synthetic set of data were used. This chapter is structured as follows. Section 4.2 describes the essential aspects of classic analytic model for aircraft guiding control. This section is focused on the most frequently used method as well as the most general:

the proportional navigation. In Sect. 4.3, the system is defined. The state parameters and the relationship between them are also described. Section 4.4 is organized in two main parts. In the first part, the laws of motion in guiding process are couched in terms of fuzzy logic. The second part describes the steps of the proposed fuzzy model synthesis and the generic architecture of the fuzzy controller. Section 4.5 is dedicated to the controller's testing by simulation. The outputs of fuzzy system and from the other particular analytic models are comparatively run for different scenario of target moving. Section 4.6 discusses the results and their impact in practice. The conclusion and future work are explained in Sect. 4.7.

4.2 The Guiding Issue at a Glance

The navigation trajectory is generally established according to the aircraft designation to the characteristics of the flight environment and to certain conditions set on the flight mission. A particular case of air navigation is that of guiding aircrafts in special missions. In the main, the navigation problem is the same, but the trajectory is dictated by conditions that require destruction of a target or by the requirement that the aircraft should precisely reach a point in space at the certain time. In addition there are any cases of imposed boundaries or ad-hoc restrictions affecting a possible trajectory. Under such circumstances, the necessary trajectories are complex and difficult to predict and the evolution control must meet the requirements of real time, robustness and flexibility according to the contemporary air missions. From this point of view, on a smaller scale, we may consider the deterministic and stochastic differential classical mathematical modeling.

The proportional guidance is a very popular method in flight to target interception control. Large amounts of literature are dedicated to this issue, but one of the best I consider is the monograph [7]. The basic schema for a typical target interception by a guided missile is shown in Fig. 4.1. There are few parameters that describe the state of the system in this case. The guiding principle consists in an imposed analytic proportional relationship between two cinematic parameters of tracing process: the angular speed of tangential path – $\dot{\theta}$, and the angular speed of sight line to target – $\dot{\Psi}$, as follows:

$$\dot{\theta} = K \cdot \dot{\Psi} \tag{4.1}$$

where K is the proportional guiding constant. The angular velocity of sight line is described by an algebraic relationship based on the geometry of the process and the magnitudes of the velocity vectors of the two mobile objects, as follows:

$$\dot{\Psi} = v_V \cdot \sin(u_V - \beta) - v_T \cdot \cos(\beta - \alpha) \tag{4.2}$$

The angular parameters in the above equation are derived from the following relationships: $u_V = \arctan(\dot{y_V}/\dot{x_V})$, $\beta = \arctan(\Delta y/\Delta x)$ and

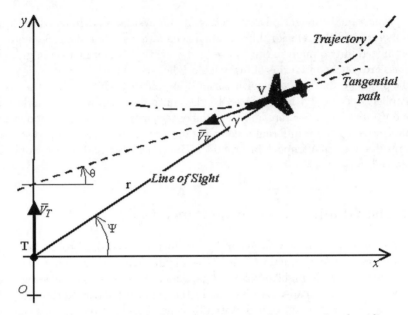

Fig. 4.1. The geometry of target interception by proportional guiding

$\alpha = \arctan(\dot{y}_T / \dot{x}_T)$. The (2.2) is useful in simulation. If the velocity profiles are given we derive the angular speed of sight line, and then become possible to solve the differential Equation (2.1) in order to obtain the trajectory. Preliminary, the initial states for integration have to be chosen. In this case, an exact integration could be done only under very limitative circumstances regarding velocity profiles: $v_V = const$ and $v_T = const$. Obviously, this is a very unlikely case in practice therefore the problem can be generally solved by a numerical integration method. We try to give up using rigid, time-consuming analytical models sensitive to disturbances and susceptible to accumulate errors. Thus we find more suitable to replace the traditional models with an alternative based on logical representations and heuristic descriptions.

4.3 The Guiding Control System

The fuzzy approach of the guiding process aims to describe a guiding cinematic law, producing evolution of aircraft state parameters in order to fulfill the flight task. As you can very well see in Fig. 4.2, an additional loop appears in the control circuit which contains the mission block as a reference model, which is in fact the flight program (law of guiding). By adopting a Cartesian coordinates ground reference system, see Fig. 4.3, in order to define the state of the guided aircraft V and the target T, we consider the position and mass centre displacement variable parameters for the two mobile objects as the 6-dimensional vectors as is following:

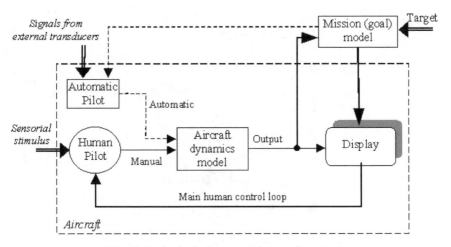

Fig. 4.2. A block diagram of aircraft control

$$S_V = [x_V\, y_V\, z_V\, \dot{x}_V\, \dot{y}_V\, \dot{z}_V]^T \qquad (4.3)$$

$$S_T = [x_T\, y_T\, z_T\, \dot{x}_T\, \dot{y}_T\, \dot{z}_T]^T \qquad (4.4)$$

The control data are established by choosing the output variables in the state vector (2.3) in terms of the guided aircraft velocity components:

$$V_C = [\dot{x}_V\, \dot{y}_V\, \dot{z}_V]^T \qquad (4.5)$$

We consider the motion conventional decomposed into the two reference planes: the horizontal one, and the other vertical. The horizontal plane is defined by xOy elements while the vertical plane is defined as follows: it contains the two points V and T and it is always parallel with the vertical axis Oz. We have to note the horizontal plane is fixed while the vertical plane generally has a lot of different instantaneous positions. As it is considered the particular situation in Fig. 4.3, we introduce the supplementary state parameters by the notations as follows:

$$\Delta x = x_T - x_V \qquad (4.6)$$

$$\Delta y = y_T - y_V \qquad (4.7)$$

$$\Delta z = z_T - z_V \qquad (4.8)$$

$$r = \sqrt{\Delta x^2 + \Delta y^2} \qquad (4.9)$$

The variable r denotes the distance between the aircraft and target in the horizontal plane. Additionally, we ought to note that distance r belongs to the trace of vertical plane on the horizontal plane.

The presented arrangement of the system has the few advantages in practice. First, it stands out the set of state parameters easily acquired by measurements from sensors or from the other ground-based information systems.

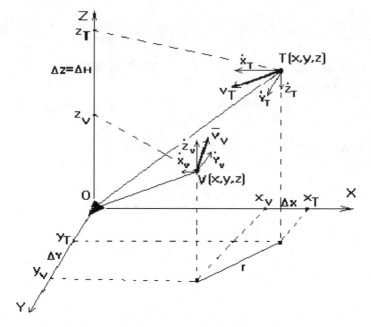

Fig. 4.3. The reference system for kinematics process

Second, despite the fact the Cartesian coordinates are not usual in global navigation the system can be easily converted to other one. Last but not least, the state parameters highlighted in the system are very suitable to be interpreted and transposed in fuzzy rule based knowledge.

4.4 Fuzzy Model

In the context of contemporary aero spatial flights, navigation systems play an essential role in the success of mission. Navigation is the task engaging the most hardware and software resources as well as more human experience and energy during a flight mission. Despite automation, generally flight navigation still needs to be strongly assisted by human pilot and/or flight controller. As a matter of fact, the concept of human like thinking machine should be implemented in benefit of systems efficiency. There are important reasons which have led us to develop a fuzzy approaching in air navigation control. First is a problem of uncertainty of the model as well as of the measured parameters in the system. On one hand, any model can not perfect describe a process. Consequently, the existing models are usually not accurate enough to fulfill the description of specific situation. On the other hand, the measurements of the parameters are not accurate because of the instruments tolerance. Second, the high nonlinear dynamics of flight makes the analytical based models very sensitive on errors. Typical features of control systems involve considerable complexities related with their behavior. Third, the specific situations of guided flight require strong real time capabilities which are difficult to meet with the cronophag analytic models. Generally, in this case the models are differential and need the iterative solving techniques. All in all, the need of flexibility and easy customing for various flight tasks needs a maintainable model and real learning and adapting properties. The guided aircraft evolution can be fuzzy described considering a fuzzy rules based approximate reasoning in the context of a decision making process. In Fig. 4.4 the chart of this process is presented.

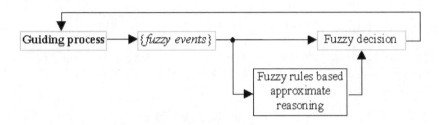

Fig. 4.4. Guiding process as a fuzzy decision

A fuzzy event is understood as any change of the flight dynamics in guiding process. The fuzzy rules based systems manages the fuzzy events providing a fuzzy decision that finally counts in guiding process. Considering the cinematic parameters of the two mobile objects as the fuzzy variables we can establish the guided aircraft motion laws as the generic functions as follows:

- in the horizontal plane:

$$\dot{x_V} = f_x(\dot{x_T}, \Delta x, r) \tag{4.10}$$

$$\dot{y_V} = f_y(\dot{y_T}, \Delta y, r) \tag{4.11}$$

- on the vertically:

$$\dot{z_V} = f_z(\dot{z_T}, \Delta z, r) \tag{4.12}$$

4.4.1 Fuzzy Model Design

The main steps in fuzzy model designing follow the standard phases of fuzzy systems development as is following:

1. Settling the relevant variables which describe the phenomena: as the fuzzy inputs and the fuzzy outputs.
2. Establishing the universe of discourse of the variables.

3. Defining the fuzzy sets and their associated membership functions for each variable in close relationship with an appropriate linguistic description of the process.
4. If-Then rules generation and choosing the related issues: the connection of the antecedents and the weight of each rule.
5. Adopting the methods involved in the fuzzy inference system: T-norm, S-norm, implication and aggregation methods as well as the defuzification technique. Considering that the reader is familiar with basics of fuzzy systems we follow the above described path focused on the application. The formal part of the steps also will not be more expounded than the application requires.

Fuzzy Variables

The generic model of cinematic in the aircraft approaching the target process, contains three discrete functional equations in the time domain for displacing the weight point as is following:

$$\dot{x}_V(t_{k+1}) = f_x[\dot{x}_T(t_k), \Delta x(t_k), r(t_k)] \tag{4.13}$$

$$\dot{y}_V(t_{k+1}) = f_y[\dot{y}_T(t_k), \Delta y(t_k), r(t_k)] \tag{4.14}$$

$$\dot{z}_V(t_{k+1}) = f_z[\dot{z}_T(t_k), \Delta z(t_k), r(t_k)] \tag{4.15}$$

These relationships describe the generic sequential dependencies between the process variables and the state variables as they are acquired from the process at the instants $t_k, k = 1, \ldots, n$, into a finite time interval. In the formal description given by (2.13)–(2.15) the process variables $\dot{x}_V, \dot{y}_V, \dot{z}_V$ are considered outputs while the state variables as arguments of the functions are considered inputs. The universe of discourse of each fuzzy variable is limited to maximum pertinent values. As a matter of fact, we have adopted the scalable domains. The scaling of each axis is made with an appropriate factor depending on the meaning of variable. In Fig. 4.5, the scaling factors are highlighted on each abscissa of the graphics.

Fuzzy Sets and Linguistic Attributes

A number of associated fuzzy sets can be intuitively chosen for each variable on its universe of discourse. The fuzzy sets are described by free chosen membership functions and are intuitively denominated by the linguistic attributes. Each variable is described by a term set consisting of a finite numbers of linguistically expressed values (attributes). Thus, for the inputs $\Delta x, \Delta y, \Delta z$ and for the outputs $\dot{x}_V, \dot{y}_V, \dot{z}_V$ seven fuzzy sets are defined, three fuzzy sets are associated to the input variable r and five fuzzy sets are defined for the inputs $\dot{x}_T, \dot{y}_T, \dot{z}_T$. The membership functions were accordingly settled regarding their shape and their relative distribution corresponding to an axis. Few aspects have to discuss related this issue. First, choosing the membership function

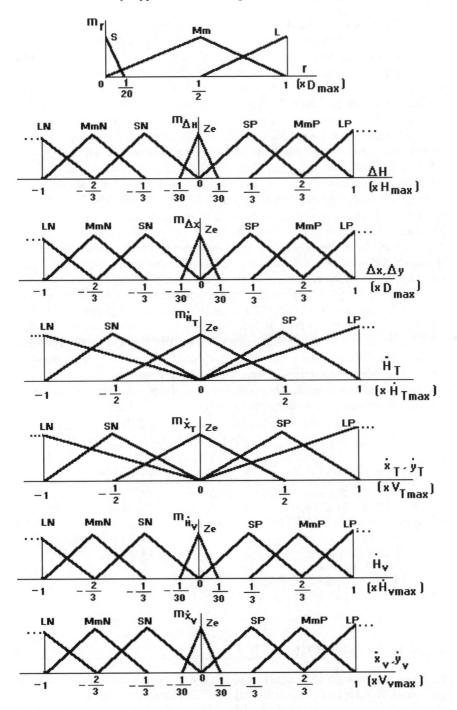

Fig. 4.5. The membership functions

is one of the most disputed issues of fuzzy systems theory. Second, choosing the membership functions are relevant with respect to how they influence the fuzzy controller performance, as the following: speed of response, continuities of the output, accuracy of control, etc.. Third, in practice is currently use few criteria that guarantee the best performance of a fuzzy controller. An essential criterion has been proven to be the cross-point ratio between two membership functions. This parameter defines the cross-point level and the overlapping degree of two membership functions. Generally, the choice of cross-point level of 0.5 and a cross-point ratio of 1 (i.e. one cross-point) is usually reported in the literature. Regarding the shape of the membership functions there are very few alternatives in practice. In fact, there are two classes of shapes: linear (triangular, trapezoidal) and curvilinear (Gaussian, sigmoid, etc.). The shape of the membership functions does not play a significant role, but is notable that trapezoidal functions are responsible for a slower rise time of the system's response [2]. In the previous work [5] have been proved the low sensitivity of a fuzzy control system for different shapes of membership functions. In order to improve the performance of the fuzzy control systems the membership functions are subject of the adaptively mechanisms. The membership functions are illustrated in Fig. 4.5. Their shapes are linear in order to reduce as possible the computing effort. It is a notable manner of defining the membership functions supports as the certain numerical ratios. The main advantage of this definition consists of the constancy of the fuzzy sets overlapping degrees. The linguistic attributes were suggestively chosen in order to describe the process as meaningful. These are given in Table 4.1 which also contains the abbreviations used in the next.

Table 4.1. The assigned linguistic values of fuzzy variables

$\dot{x}_V, \dot{y}_V, \dot{z}_V, \Delta x, \Delta y, \Delta z$	r	$\dot{x}_T, \dot{y}_T, \dot{z}_T$
LN: Large Negative	S: Small	LN: Large Negative
MmN: Medium Negative	Mm: Medium	SN: Small Negative
SN: Small Negative	L: Large	Ze: Zero
Ze: Zero		SP: Small Positive
SP: Small Positive		LP: Large Positive
MmP: Medium Positive		
LP: Large Positive		

Fuzzy Rules Base (FRB)

FRBs are achieved by logically connecting the fuzzy sets associated to the output variables with fuzzy sets of the inputs. The control strategy is expressed in linguistic terms on which basis the logic inferences are formed that will make the FRB's rules. Practically, a rule appears when there is a premise referring to an event implying or causing a certain consequence. In general, any physical process may be modeled based on its description by rules. This means establishing a set of antecedents (premises) and identification of the consequences set (in fact, there is an identification of a cause-effect relationships set). As a result, a fuzzy rule is formulated by adding the $Pi(i = 1, \ldots, n)$ fuzzy propositions (antecedents) with the help of acknowledged logical operators (generically noted with the symbol $\bigotimes$) and by equalizing the result to the C consequence, that is:

$$P1 \bigotimes P2 \bigotimes \cdots \bigotimes Pn \Rightarrow C \qquad (4.16)$$

Guiding Strategy in the Horizontal Plane (GHP)

GHP is a navigation process to achieve the direction that should be followed by the aircraft to the moving target. Usually, this task is solved based on the predicted alignments where the target could be meeting. Under the present circumstances the guiding operation is characterized by no deterministic mathematical relationships among alignments. The non-determination degree in this model is directly proportional to increasing the evolution possibilities in time and space of contemporary aircrafts. The alignments are approximately defined (necessary, possible) and depend on the target evolution hypotheses. The GHP strategy is expressed in logical statements (IF-THEN) at the working fuzzy variables, which correspond to the two orthogonal directions chosen in the ground plane as is following:

- On the Ox axis direction: IF $\dot{x}_T = $ <linguistic term> and $\Delta x = $ <linguistic term> and r = <linguistic term> THEN $\dot{x}_V = $ <linguistic term>.
- On the Oy axis direction: IF $\dot{y}_T = $ <linguistic term> and $\Delta y = $ <linguistic term> and r = <linguistic term> THEN $\dot{y}_V = $ <linguistic term>.

Therefore, by making combinations among the previous and later fuzzy sets with the logical conjunction operation (AND) we get the guiding rules. As the fuzzy variables corresponding to the two command channels (the Ox, Oy directions) have identical fuzzy sets, a unique FRB can be designed for the two motions. For the particular case of a moving target interception operation, the logic could have the following description:
"The aircraft makes for the target the faster:

- the greater the target speed to the objective;
- the greater the distance between the aircraft and the target;
- the smaller the distance between the target and its destination"

Starting from this form of raising the problem we can naturally get to information structured under the form of the rules. Corresponding to the number of linguistic terms attributed to the fuzzy sets for each variable, the total number of logical statements for each command channel is in this case: $(7 \times 5 \times 3) = 105$. FRB admits of three-dimensional geometric representation under the form of a parallelepiped divided into 105 elementary cells, each attributed to a particular rule. In Fig. 4.6 shows the cellular representation and in Fig. 4.7 an unfolded completed form with rules for GHP strategy is illustrated.

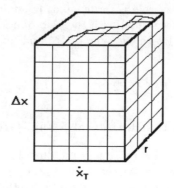

Fig. 4.6. Generic view of FRB

$$\dot{x}_T \cdot \dot{y}_T \qquad\qquad \dot{x}_T \cdot \dot{y}_T \qquad\qquad \dot{x}_T \cdot \dot{y}_T$$

$\Delta x / \Delta y$, $r = s$:

	LN	SN	Ze	SP	LP
LN	LN	LN	LN	LN	LN
MmN	LN	MmN	MmN	MmN	MmN
SN	MmN	SN	SN	Ze	Ze
Ze	NMi	Ze	Ze	Ze	SP
SP	Ze	Ze	SP	SP	MmP
MmP	SP	SP	MmP	LP	LP
LP	LP	LP	LP	LP	LP

$\Delta x / \Delta y$, $r = Mm$:

	LN	SN	Ze	SP	LP
LN	LN	LN	LN	LN	LN
MmN	LN	LN	LN	LN	LN
SN	LN	LN	MmN	LN	LN
Ze	MmN	MmN	SN	MmN	MmN
SP	LP	LP	MmP	LP	LP
MmP	LP	LP	LP	LP	LP
LP	LP	LP	LP	LP	LP

$\Delta x / \Delta y$, $r = L$:

	LN	SN	Ze	SP	LP
LN	LN	LN	LN	LN	LN
MmN	LN	LN	LN	LN	LN
SN	LN	LN	LN	LN	LN
Ze	LN	MmN	MmN	MmN	LN
SP	LP	LP	LP	LP	LP
MmP	LP	LP	LP	LP	LP
LP	LP	LP	LP	LP	LP

Fig. 4.7. A particular FRB for GHP strategy

Vertical Guiding Strategy (VG)

The flight profile is defined by the vertical aircraft evolution according to the ground position corresponding to any given flight moment. The flight profile command strategy is exclusively based on both technical and tactical considerations. Usually, this stage of guiding is little conditioned by the aircraft evolution in the horizontal plane, so that VG strategy could be performed in an independent designing stage. There are possible a lot of strategies to vertically guide the aircraft, but the main choice criteria are technical: the fuel consumption, respectively tactical: anti-foe protection or obstacles avoidance.

The strategy chosen for controlling motion on the Oz axis is expressed by logical inferences similar to horizontally guiding as is following:

IF $\dot{z}_T$ = <linguistic term> and Δz = <linguistic term> and r = <linguistic term> THEN $\dot{z}_V$ = <linguistic term>

The total number of fuzzy rules in a FRB for VG is also 105 but their content is specific to the flight profile strategy.

4.4.2 Fuzzy Algorithm

The fuzzy systems process information according to an adequate philosophy that principally, has the following flow: *inputs* ⇒ *(fuzzification)* ⇒ *(inferences)* ⇒ *(defuzzification)* ⇒ *outputs*.

Each loop in the information processing chain may be achieved by different techniques and procedures acknowledged in the literature [2, 6].

Fuzzification

This operation is performed by applying the membership function corresponding to each input variable, and then it is associated with a vector whose elements are the membership degrees to the fuzzy sets defined on its domain. Thus, for the working variables values considered at a certain time t^*, the fuzzyfied variables vectors are:

$$\dot{x}_T^* \rightarrow [m_{LN}(\dot{x}_T^*), m_{SN}(\dot{x}_T^*), m_{Ze}(\dot{x}_T^*), m_{SP}(\dot{x}_T^*), m_{LP}(\dot{x}_T^*)] \quad (4.17)$$

$$\Delta x^* \rightarrow [m_{LN}(\Delta x^*), m_{MmN}(\Delta x^*), m_{SN}(\Delta x^*), m_{Ze}(\Delta x^*),$$
$$m_{SP}(\Delta x^*), m_{MmP}(\Delta x^*), m_{LP}(\Delta x^*)] \quad (4.18)$$

$$r^* \rightarrow [m_S(r^*), m_{Mm}(r^*), m_L(r^*)] \quad (4.19)$$

In the same manner similar relationships for variables y and z are derived. In the expressions (2.17), (2.18) and (2.19) the generic form $m_A(B)$ denominates the membership degree of B in fuzzy set A, where $m_A(B) \in [0, 1]$.

Inference

Each moment of time t_*, the fuzzy algorithm activates the rules in FRB as a parallel process. Each rule output is a fuzzy value. Each rule into FRB represents a logical expression developed with logical operator AND. We adopt a T-norm in form of well-known Zadeh's relationship based on the minimum operation. Therefore, the fuzzy sets intersection operation is applied at whose output we get a punctual minimum of the membership functions on the whole definition domain of the output variables. Working with Zadeh's fuzzy T-norm [2], after the inference on a generic rule we get the following degree of activation of the $i - th$ rule in FRB, $(i = 1, \ldots, r)$:

$$w_i = \min[m(\dot{x}_T^*), m(\Delta x^*), m(r^*)]. \qquad (4.20)$$

Usually it is said that the rule i is faired with a degree of activation equal by the value $w_i \in [0, 1]$. For the variables y, z, we reason in the same manner and the obtained relationships are similar. We note that here the individual-rule based inference is applied [2]. In this method two steps are performed: first each single rule is fired computing the degree of activation on the antecedents with (2.20), and second, each rule-consequent is scaled by the computed degree. The form of each fuzzy set activated on the whole universe of discourse of the output variable depends on the "encoding" scheme used. We adopt a specific encoding procedure [6], according to which the global fuzzy system output results from the output variable membership functions weighted by the activation degree of the i rule. In terms of the membership functions this is expressed as:

$$m_i(\dot{x}_V^*) = wi \cdot m_i(\dot{x}_V^*); \qquad (4.21)$$

Basically, the effect of this operation is a weighting while in a geometrical sense it acts as if the output's membership function is reshaped. For the y and z variables, the relationships are similar. This operation mode is acknowledged in literature as the correlation-product inference [6]. These operations performed so far are structurally illustrated in Fig. 4.8. We notice that the rules used in the inference process may have as a result the same fuzzy output sets as consequents, generally scaled by different w_i degrees. The inference operation is therefore completed at the level of the whole FRB by an aggregation (composition) technique of the primary inference results (from each i rule activated). In this application we adopted an aggregation method known as MAX rule, according to which, for the fuzzy rules having the same output set as consequent, it is engaged with the maximum value of the w_i degree. This calculus sequence is applied for all basic rules and similarly for each group of variables corresponding to the motion Equations [relationships (2.13), (2.14) and (2.15)]. At this stage of the algorithm each output is still a fuzzy parameter.

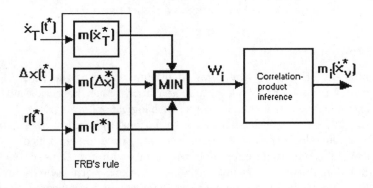

Fig. 4.8. The fuzzy inference and global rules composition

After engaging all the FRB's rules at the time t^*, the result is generated by the fuzzy sets associated with the output variables:

$$x_V^{\cdot *} \rightarrow [m_{LN}(x_V^{\cdot *}), m_{MmN}(x_V^{\cdot *}), m_{SN}(x_V^{\cdot *}), m_{Ze}(x_V^{\cdot *}), m_{SP}(x_V^{\cdot *}),$$
$$m_{MmP}(x_V^{\cdot *}), m_{LP}(x_V^{\cdot *})] \qquad (4.22)$$

For the y and z outputs the similar sets are also obtained. Therefore, the output variable amount is represented by a fuzzy set O, obtained by adding all mi scaled fuzzy sets generated on the respective variable domain, going through the whole FRB (see Fig. 4.9):

$$O = \sum_{i=1}^{r} w_i \cdot m_i \qquad (4.23)$$

Defuzzification

This operation consists in taking a deterministic scalar value out of the fuzzy information associated with the output variable under the form of O set. From the best known defuzzification methods [2] we have chosen the weight point method (centroid). According to it, the output significant value v_k is calculated as the weight point coordinate value of the plane domain designated by the areas sum of the elementary fuzzy sets – $m_i(x)$, on the x real axis that defines the universe of discourse of the respective output variable. Calculus of the plane figures weight point resulting from superposing different fuzzy sets raises some problems. Generally, the discrete variant is used for which the acknowledged methods of digital integration are applied. Yet, in case of multiple variables fuzzy models working with a great number of rules and/or having more complicated membership forms, the process becomes toilsome. In order to reduce the calculation time, to simplify the numerical algorithm and to implement hardware in defuzzification circuits, a discrete method will be used, based on a theorem stated and demonstrated in [6], according to which is the following relationship:

$$v_k = \frac{\sum_{i=1}^{r} w_i \cdot c_i \cdot I_i}{\sum_{i=1}^{r} w_i \cdot I_i} \qquad (4.24)$$

where I_i and c_i define the surface area and the coordinate of output fuzzy set weight point respectively, corresponding to the i-th rule, $(i = 1, \ldots, r)$. Replacing by defuzzification the fuzzy system output becomes a deterministic value of the fuzzy control system output variable $x_V^{\cdot}$ at the time t^*. Therefore the diagram in Fig. 4.7 is extended for the whole FRB, as you can see in Fig. 4.8.

4.4.3 Fuzzy Controller Structure

Configuration of the fuzzy controller designed to guide the aircraft 3-dimensionally takes into account the conventional decomposition of the motions corresponding to the evolution strategies adopted in the guiding systems. From

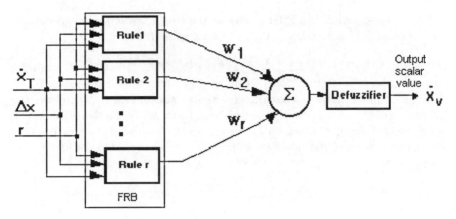

Fig. 4.9. Global FRB's output and defuzzification

the point of view of the dynamics of flight, the aircraft motion is approached conventionally into two types of coupled motions: the longitudinal motion, respectively lateral motion. The dynamic coupling of the two motions appears by means of the so called coupling terms in the mathematical model at the motion equations level around the weight point. The fuzzy controller has a three-channel structure to find the weight point motion command variables on the reference orthogonal system axes. The command channels are quasi-independent, meaning there is an information coupling that requires a one-way dependence from the point of view of command itself. This comes from the fact, already mentioned, that guiding in the horizontal plane is independent of the flight profile guiding while the latter depends on the evolution in the horizontal plane. As a matter of fact, this interdependence is required by the methodology of aircraft guiding process too, taking it into account during the rule bases working out phase. Figure 4.10 shows the fuzzy controller suggested variant, designed to guide the aircraft in the two planes. The inputs on the left side slots are acquired at the defined moments of time $t_k, k = 1, \ldots, n$, while the outputs are obtained at the t_{k+1} , according to the model of the process generically described by (2.13)–(2.15). The additional integrating blocks in the controller's diagram provides the coordinates useful in instrumental navigation.

4.5 Simulation

A main stage in validating the response of fuzzy controller is given by simulating its model comparatively with the proportional guiding model taken as a reference. The simulation is based on synthetic data reflecting realistic conditions for the process considered. The response of fuzzy guiding model (FM) will be compared with the response of proportional navigation model (PN)

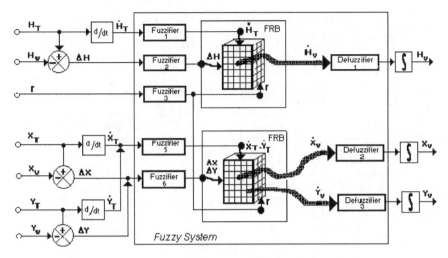

Fig. 4.10. The block diagram of navigation fuzzy controller

under three particular hypotheses regarding the speed of interceptor. The versions of the PN model are based on the following considered hypotheses:

- The speed of interceptor is given as $v_V = constant$.
- The speed of interceptor is quite equal with speed of target $v_V = v_T$.
- The interceptor adopts the speed calculated by fuzzy model $v_V = v_{Vfuzzy}$.

In the following, we refer shortly and graphically to above mentioned models as follows: $PN_{v_V=const}$ (magenta), $PN_{v_V=v_T}$ (red), $PN_{v_V=v_{Vfuzzy}}$ (green). The fuzzy guiding model is referred as FM and its related responses are outlined in blue. The target is referred as T and the related evolution of its parameters are outlined in gray. The next point is to choice the target moving scenario. This study shows the response simulation results concerning the x and y control channels for both target motion scenario: a trajectory generated by points (Case A) and a sine trajectory suggesting a highly maneuverable target respectively (Case B). In Case A, the target evolves on a trajectory that is generated point by point as a discrete model such as the following:

$$x_T = x_T(t_k), y_T = y_T(t_k), z_T = z_T(t_k) \qquad (4.25)$$

where t_k is the discrete time when the data are acquired. The states of target in case A are given as if they provide from measurements during five minutes on process and they are given in Table 4.2.

The cinematic parameters of target in case B are given by the following discrete model:

$$\begin{aligned} x_T &= x_{T0} - a \cdot t_k \\ y_T &= y_{T0} + b \cdot \sin(\Omega \cdot t_k) \\ z_T &= z_T(t_k) \end{aligned} \qquad (4.26)$$

Table 4.2. A synthetic data set of target position

No. Points k	Time $t_k[s]$	Trajectory Coordinates $x_T[m]$	$y_T[m]$	Altitude $z_T[m]$
1	0	75,000	75,000	5,000
2	10	71,000	75,000	5,000
3	20	65,000	69,000	4,000
4	30	62,000	66,000	3,000
5	40	60,000	64,000	2,500
6	50	57,000	60,000	2,300
7	60	55,000	57,000	2,000
8	70	53,000	53,000	1,500
9	80	51,000	50,000	1,400
10	90	50,000	49,000	1,200
11	100	49,000	47,000	1,000
12	110	48,000	45,000	800
13	120	47,000	43,000	600
14	130	46,000	40,000	500
15	140	45,000	39,000	400
16	150	44,000	38,000	300
17	160	43,000	37,000	200
18	170	42,000	35,000	200
19	180	40,000	33,000	150
20	190	39,000	32,000	150
21	200	38,000	31,000	200
22	210	37,000	30,000	200
23	220	35,000	28,000	250
24	230	34,000	27,000	250
25	240	33,000	26,000	300
26	250	32,000	25,000	300
27	260	30,000	25,000	300
28	270	28,000	25,000	300
29	280	27,000	25,000	300
30	290	25,000	25,000	300

where a and b are the parameters of laws of motion, Ω is the angular speed of sine motion. The initial states for two simulation cases are given in Table 4.3. The results concerning the speed component of the interceptor: $\dot{x}_V, \dot{y}_V$ and the current position of the interceptor given by the coordinate's x_V, y_V are had shown comparatively in Figs. 4.11 to 4.14 and Figs. 4.15 to 4.18 respectively. Different scenarios based cases could be imagined mainly regarding the target moving model with the particular initial points. More details about the simulation algorithm as well as the programs sources are given in [3].

Table 4.3. Initial states and specific values

Parameter	Case A	Case B
Initial Position of the interceptor	$x_{V0} = 10{,}000m$ $y_{V0} = 100{,}000m$ $z_{V0} = 1{,}000m$ $\Delta t = t_k - t_{k-1} = 10s$	$x_{V0} = 0m$ $y_{V0} = 0m$ $z_{V0} = 10{,}000m$ $\Delta t = 3s$
Initial Position of the target	$x_{T0} = 75{,}000m$ $y_{T0} = 75{,}000m$ $z_{T0} = 5{,}000m$	$x_{T0} = 100{,}000m$ $y_{T0} = 80{,}000m$ $z_{T0} = 5{,}000m$
Initial speed of the interceptor (absolute values)	$\dot{x}_{V0} = 400m/s$ $\dot{y}_{V0} = 300m/s$	$\dot{x}_{V0} = 400m/s$ $\dot{y}_{V0} = 300m/s$
Initial speed of the target (absolute values)	$\dot{x}_{T0} = 400m/s$ $\dot{y}_{T0} = 0m/s$	$\dot{x}_{T0} = 300m/s$ $(\dot{y}_{T0})_{max} = b = 200m/s$
Proportional guiding parameters	$K = 3.5$ $\gamma_0 = 0°$	$K = 3.5$ $\gamma_0 = 0°$

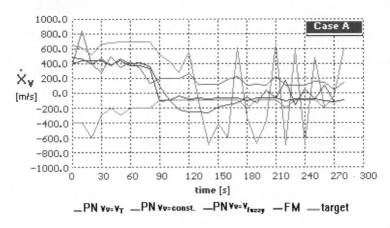

Fig. 4.11. The controlled speed component $\dot{x}_V$ diagram. Case A

4.6 Discussion

Two main aspects of the response have to be taken into consideration: the qualitative and the quantitative. The results are analyzed according to few criteria taking into consideration the capability of methods to trace the target. Several particular criteria could be formulated regarding the interception flight tasks as the following: the speed tracking capability, the capability to perform the approaching in to an acceptable, stable proximity of the target, the smoothness of curves that describe the evolution of cinematic parameters (especial for speeds), minimal time control, etc. Performing a detailed analyzes on the responses drawn in different colors (in conformity with the legend

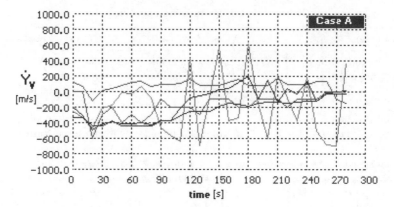

Fig. 4.12. The controlled speed component $\dot{y}_V$ diagram. Case A

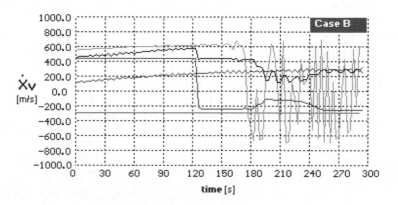

Fig. 4.13. The controlled speed component $\dot{x}_V$ diagram. Case B

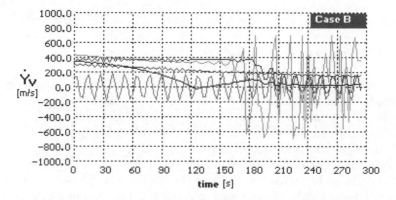

Fig. 4.14. The controlled speed component $\dot{y}_V$ diagram. Case B

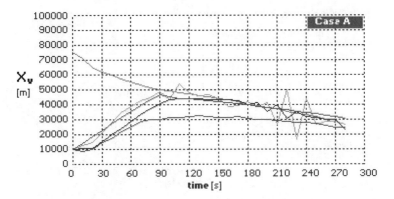

Fig. 4.15. The controlled coordinate x_V diagram. Case A

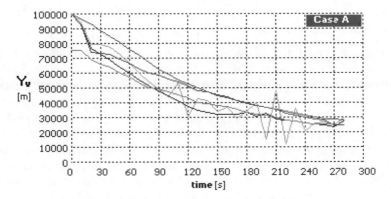

Fig. 4.16. The controlled coordinate y_V diagram. Case A

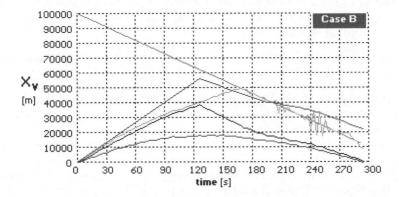

Fig. 4.17. The controlled coordinate x_V diagram. Case B

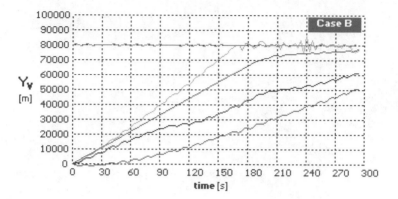

Fig. 4.18. The controlled coordinate y_V diagram. Case B

of Fig. 4.11) in relation with target curve (colored in gray) we extract useful conclusions as the following:

- In Fig. 4.11 and 4.12, Case A, the speed responses for different models of guidance could be ranked from the worst case that is the proportional navigation when the speed of interceptor is constant $v_V = const$ (magenta curve), to the best case that is of fuzzy model (blue curve). The similar situation we have in Case B. The intermediate cases are the proportional navigation when the speed of interceptor is quite equal with speed of target $v_V = v_T$ (red curve) and the proportional navigation when the interceptor adopts the speed calculated by fuzzy model $v_V = v_{Vfuzzy}$ (green curve). This latest case is quite interesting because it was derived by mixing the analytic relationship given by (1) with fuzzy rule based technique for interceptor's speed. However, this combined technique provides the better speed responses than any other version of proportional method.
- In Fig. 4.15 to 4.18 the evolution of coordinates is comparatively presented. The best tracking capability belongs to fuzzy model in both cases A and B (according to blue curve).

The fuzzy controller response is globally characterized from the qualitative point of view by the following features:

- Speed tracking capability is quite good. Observing the blue colored curves the diagrams of Fig. 4.11 to 4.14, we can see the response steady tendency to get closer to and to superpose on the evolution of the target motion parameters. The results are quite good in case of high maneuverable target (Case B).
- Notice the steady and consistent closeness of the moving objects coordinates (see diagrams in Fig. 4.15 to 4.18), in the limits of an acceptable proximity according to the control strategy implemented into the controller by FRB.

The main advantages revealed is that, in case of the models based on the kinematical parameters fuzzy process, the response is consistent and steady, meaning it has a robust behavior. On one hand, the fuzzy model gives tolerant responses to the disturbances appearing in the process kinematics. Even for non-uniform target evolutions, with great non-linearity of the motion parameters, to an entirely random, chaotic motion (a hypothetical case), the fuzzy response positive characteristics are maintained [3]. On the other hand, the compared analysis performed here proves that the flight methods based on analytical models possess a higher degree of unsteadiness, being susceptible to the guiding process dynamic disturbances. Despite its generality and flexibility greater than with other methods, the case of proportional navigation model, chosen as reference in the present work is still tributary to the analytical relatively rigid model describing the guiding law. The quantitative result interpretation depends on the simulated situation context and on the fuzzy controller functional features. Thus, using the two main adapting mechanisms and/or the fuzzy rules bases adapting mechanism respectively, we can obtain a fuzzy system which provides optimum responses, valid for a given case multitude, as to certain required technical-tactical criteria.

4.7 Summary

In this chapter a fuzzy logic based method for air navigation in interception flight has been developed. A three-dimensional cinematic model of two high maneuverable aircrafts has been considered to describe the guiding in terms of a Mamdani rule based system. Using the specific procedures of fuzzy information processing, an intelligent controller's architecture has been proposed. The response of the fuzzy controller has been simulated with the synthetic data and compared with few particular forms of the traditional proportional navigation method. The results are very promising in order to be integrated in the future Command, Control, Computer, Communication and Intelligence (C^4I) systems. In a similar way the fuzzy logic based technique can be used at making the response, practically for every parameter of the flight control process considered as an output variable. Thus, we can work out control functions for the aircraft's actuators (aerodynamic command surfaces, gas dynamic elements, etc), for the engine traction control and the laws of acting on the control parameters control equipment (stick, rudder, gas lever, etc.) by the human pilot. Starting from these possibilities, in the future research we can design various types of controllers with interesting applications to:

- Unmanned air vehicles;
- Take off/landing automatic system [4];
- Aircraft control during the flight refuel phase;
- Spacecrafts on the orbit coupling control.

First, an important stage in the development of the designed fuzzy control system consists in implementing software, fuzzy control program and developing specialized hardware structure to fuzzy process information as embedded systems. Second, the development perspectives of this work rely on the improving of the main fuzzy model's elements: the membership functions and the fuzzy rules. Basically this task is related on the fuzzy designing methodology. In a way, we have established the fuzzy sets associated with the working variables by free adopting the membership functions. Generally, the form of the membership functions, as well as the distribution of the fuzzy levels, expressed in linguistic values is a problem referring to the process of analyst's experience and intuition. This issue is generally solved by trials to get the most advantageous response of the fuzzy system as to an imposed criterion. As a result, the membership functions will be defined in a general, flexible form to allow for their adaptation according to the context. All in all, the fusion of certain information approach techniques based on genetic algorithms with the approaches already acknowledged in the field of fuzzy and neuronal systems will lead to high level artificial intelligence avionics systems.

References

1. St. J. Andriole, Advanced technology for command and control systems engineering, AFCEA International Press, Fairfax, Virginia, 1990
2. D. Driankov, H. Hellendoorn and M. Reinfrank, An Introduction to Fuzzy Control, Springer-Verlag, Berlin Heidelberg, 1993
3. S. Ionita, Contributions on optimal guiding of aircrafts towards high speed targets, (in Romanian) Doctoral Thesis, Technical Military Academy, Bucharest, Romania, 1997
4. S. Ionita, E. Sofron, The Fuzzy Model for Aircraft Landing Control, 2002 AFSS International Conference on Fuzzy Systems, Calcutta, India Feb. 3–6, 2002, in Lecture Notes in Artificial Intelligence, Springer-Verlag, 2002, pp. 47–54
5. S. Ionita, On the aspects of the membership functions shape influence in a fuzzy control model, (in Romanian), Military Technique-Scientific Supplement, No. 1, 1995, pp. 9–16
6. B. Kosko, Neural Networks and Fuzzy Systems, Prentice Hall, Englewood Cliffs, 1992
7. M.M. Nita, D. St. Andreescu, Roket's Flight (in Romanian), Military Publishing House, Bucuresti, 1964
8. P.S.V. Rao, Computers in Air Defence, Defence Science Journal, Vol. 37, October, 1987, pp. 507–513
9. K. Rokhsaz, J.E. Steck, Use of Neural Network in Control of High-Alpha Maneuvres, Journal of Guidance, Control, and Dynamics, Vol. 16, No. 5, Sept-Oct. 1993, pp. 934–939
10. J.E. Steck, S.N. Balakrishnan, Use of Neural Networks in Optimal Guidance, IEEE Transactions on Aerospace and Electronic Systems, Vol. 30, No. 1, January, 1994, pp. 287–293

11. H.J. Zimmermann, Concepts, Origins, Present Developments and the Future of Computational Intelligence, in Advances in Generalized Structures Approximate Reasoning, and Applications (F. Eugeni, A. Maturo, I. Tofan Editors), Performantica Press, Iasi, 2001, pp. 99–108
12. H.J. Zimmermann, Fuzzy Set Theory-and its Applications, Kluwer Academic Publisher, Boston, 1991

5

Hybrid Soft and Hard Computing Based Forex Monitoring Systems

A. Abraham

Computer Science Department, Oklahoma State University, USA
ajith.abraham@ieee.org, http://ajith.softcomputing.net

In a universe with a single currency, there would be no foreign exchange market, no foreign exchange rates, and no foreign exchange. Over the past twenty-five years, the way the market has performed those tasks has changed enormously. The need for intelligent monitoring systems has become a necessity to keep track of the complex forex market. The vast currency market is a foreign concept to the average individual. However, once it is broken down into simple terms, the average individual can begin to understand the foreign exchange market and use it as a financial instrument for future investing. In this chapter, we attempt to compare the performance of hybrid soft computing and hard computing techniques to predict the average monthly forex rates one month ahead. The soft computing models considered are a neural network trained by the scaled conjugate gradient algorithm and a neuro-fuzzy model implementing a multi-output Takagi-Sugeno fuzzy inference system. We also considered Multivariate Adaptive Regression Splines (MARS), Classification and Regression Trees (CART) and a hybrid CART-MARS technique. We considered the exchange rates of Australian dollar with respect to US dollar, Singapore dollar, New Zealand dollar, Japanese yen and United Kingdom pounds. The models were trained using 70% of the data and remaining was used for testing and validation purposes. It is observed that the proposed hybrid models could predict the forex rates more accurately most of the time than all the techniques when applied individually.

5.1 Introduction

Creating many international businesses, the globalization has made the international trade, international financial transactions and investment to rapidly grow. Globalization is followed by foreign exchange market also known as forex. The forex is defined as a change in a market value relationship between national currencies (at a particular point in time) that produces profits, or losses, for all foreign currency traders [12]. As such, it plays an important role of providing payments in between countries, transferring funds from one currency to another and determining the exchange rate.

The forex is the largest and the most liquid market in the world with a daily turnover of around 1 trillion U.S. dollars [16]. It was founded in 1973 with the deregulation of the foreign exchange rate in the USA and other developed countries. Namely, before 1973 the fixed exchange rates regime was used for global currency relationships. It was based on the Bretton Woods' agreement from 1944 with American dollar as an anchor for all free world currencies. The American dollar has been a reserve currency for the world that was based on gold standard. No other country guaranteed to exchange its currency for a gold. However, in 1960s and early 1970s the global economic crisis brought on by the worldwide inflation has shown that The United States were not able any more to meet the gold standard. With a rise of inflation more dollars became worth less, and dollars holders around the globe sought the safety of gold. As a consequence, many nations were unable to maintain the value of their currencies under the Bretton Woods regime, and the U.S. gold reserves significantly fell. Then, in 1973 the floating exchange rate system was created establishing markets' prices rule. The system is dynamic, generating greater trade and capital flows. It is expanding with rapid technological innovations. In particular, the foreign exchange market has become an over-the-counter market with traders located in the offices of major commercial banks around the world. Today, communication among traders goes on using computers, telephones, telexes, and faxes. Traders buy and sell currencies, but also they create prices. The exchange of currencies, however, is in the form of an exchange of electronic messages [3].

Most of the trading in the forex market takes places in several currencies: U.S dollar, German mark, Japanese yen, British pound sterling, Australian dollar, Canadian dollar etc. More than 80 percent of global foreign exchange transactions are still based on American dollar. There are two reasons for quoting most exchange rates against the U.S. dollar. The first has to do with simplicity to avoid enormous number of dealing markets if each currency were traded directly against each other currency. A second is to avoid the possibility of triangular arbitrage. That is, since all currencies are traded with respect to the dollar, there is only one available cross rate and no possibility of arbitrage [9]. The forex market is 24-hour market with three major centers in different part of the world: New York, London, and Tokyo. It is the busiest in the early morning New York time since banks in London and New York are simultaneously open and trading. Its center's open and close one after the other. If it is open in Tokyo and Hong Kong, it is also open in Singapore. Then if it opens in Los Angeles in the after noon, it will be also open in Sydney the next day in the morning. At present the forex market includes the participation of commercial banks around the globe, with a tendency to spread to corporate, funding and retail institutions [4].

At the forex market, traders create prices by buying and selling currencies to exporters, importers, portfolio managers, and tourists. Each currency has two prices: a bid price at which a trader is willing to buy and an offer price at which a trader is willing to sell. If being in the major money centers banks traders deal in two way prices, for both buying and selling. In market-making banks worldwide much of the trading take place by direct dealing, while the rest takes place through brokers. Today computerized services electronically match buy and sell orders using an automated brokerage terminal. As Grabbe quotes, about 85 percent of all forex trading is between market makers [9]. With the rest the forex purchases and sales are by companies engaged in trade, or tourism. Since the most trading takes place between market makers it creates a space for speculative gains and losses. However, speculation in the forex market is potentially a zero-sum game: the cumulative profits equal the cumulative losses. The operations are inter-bank transactions where a single rumor can create eruptive reactions followed by huge and often-unpredictable capital flows. Now traders play against each other instead of playing against central banks as they did when currencies were not floating [7].

Starting from 1983 there were considerable changes in the Australian forex market. Like Australia most of developed and developing countries in the world welcome foreign investors. When foreign investors get access to invest in any country's bond equities, manufacturing industries, property market and other assets then the forex market becomes affected. This affect influences everyday personal and corporate financial lives, and the economic and political fate of every country on the earth. The nature of the forex market is generally complex and volatile. The volatility or rate fluctuation depends on many factors. Some of factors include financing government deficits, changing hands of equity in companies, ownership of real estate, employment opportunities, merging and ownership of large financial corporation or companies. The major attractions to the business of forex trading are threefold, namely, high liquidity, good leverage and low cost associated with actual trading. There are, of course, many other advantages attached with the dealing of forex market once one gets involved and understands it in more details [10].

There are many ways in which traders analyze the directions of the market. Whatever the method, it is always related to the activities of the price for some periods of time in the past. The pattern in which the prices move up and down tends to repeat itself. Thus, the prediction of future price movements can be plotted out by studying the history of past price movements. Of course there are still other theories to be followed if an accurate prediction is to be expected. These theories are associated with financial jargons such as: support and resistance levels, trend lines, double bottoms and double tops, technical indicators, etc. It is well known that the forex market has its own momentum and using traditional statistical techniques based on the previous market trends and parameters, it is very difficult to predict future exchange rates.

Long-term prediction of exchange rates might help the policy makers and traders for making crucial decisions. We analyzed the average monthly foreign exchange rates for continuous 244 months starting January 1981 for exchange rates of 5 international currencies with respect to Australian dollar [2]. In this chapter, we report the comparative performance of neural network, neuro-fuzzy system, MARS [14], CART [15] and a hybrid CART-MARS approach. In Sects. 5.2 and 5.2.1, we provide some theoretical background on soft computing and the considered hard computing techniques followed by experiment setup, training and test results in Sect. 5.2.2. Some conclusions and future research directions are also provided towards the end.

5.2 Soft Computing

Soft computing was first proposed by Zadeh [17] to construct new genera-tion computationally intelligent hybrid systems consisting of neural networks, fuzzy inference system, approximate reasoning and derivative free optimiza-tion techniques. It is well known that the intelligent systems, which can pro-vide human like expertise such as domain knowledge, uncertain reasoning, and adaptation to a noisy and time varying environment, are important in tackling practical computing problems. In contrast with conventional AI tech-niques (hard computing), which only deal with precision, certainty and rigor the guiding principle of soft computing is to exploit the tolerance for im-precision, uncertainty, low solution cost, robustness, partial truth to achieve tractability and better rapport with reality.

5.2.1 Artificial Neural Networks

Artificial Neural Networks (ANNs) have been developed as generalizations of mathematical models of biological nervous systems. A neural network is char-acterized by the network architecture, the connection strength between pairs of neurons (weights), node properties, and updating rules. The updating or learning rules control weights and/or states of the processing elements (neu-rons). Normally, an objective function is defined that represents the complete status of the network, and its set of minima corresponds to different stable states of the network. It can learn by adapting its weights to changes in the surrounding environment, can handle imprecise information, and generalize from known tasks to unknown ones. Backpropagation is a gradient descent technique to minimize some error criteria E. A good choice of several parame-ters (initial weights, learning rate, momentum etc.) are required for training success and speed of the ANN. Empirical research has shown that backprop-agation algorithm often is stuck in a local minimum mainly because of the random initialization of weights. Backpropagation usually generalizes quite well to detect the global features of the input but after prolonged training the network will start to recognize individual input/output pair rather than settling for weights that generally describe the mapping for the whole training set.

In the Conjugate Gradient Algorithm (CGA) a search is performed along conjugate directions, which produces generally faster convergence than steepest descent directions. A search is made along the conjugate gradient direction to determine the step size, which will minimize the performance function along that line. A line search is performed to determine the optimal distance to move along the current search direction. Then the next search direction is determined so that it is conjugate to previous search direction. The general procedure for determining the new search direction is to combine the new steepest descent direction with the previous search direction. An important feature of the CGA is that the minimization performed in one step is not partially undone by the next, as it is the case with gradient descent methods. An important drawback of CGA is the requirement of a line search, which is computationally expensive. Moller introduced the Scaled Conjugate Gradient Algorithm (SCGA) as a way of avoiding the complicated line search procedure of conventional CGA. Detailed step-by-step description can be found in [13]. We used the scaled conjugate gradient algorithm to model forex monitoring systems.

5.2.2 Neuro-fuzzy Computing

Neuro-Fuzzy (NF) computing is a popular framework for solving complex problems [1]. We used the Adaptive Neuro Fuzzy Inference System (ANFIS) implementing a multi-output Takagi-Sugeno type Fuzzy Inference System (FIS) [11]. Figure 5.1 depicts the 6- layered architecture of multiple output ANFIS. The detailed function of each layer is as follows:

Layer-1: Each node in this layer corresponds to one linguistic label (excellent, good, etc.) to one of the input variables in the input layer $(x, \ldots, y)$. In other words, the output link represent the membership value, which specifies the degree to which an input value belongs to a fuzzy set, is calculated in this layer. A clustering algorithm (usually the grid partitioning method) will decide the initial number and type of membership functions to be allocated to each of the input variable. The final shapes of the MFs will be fine tuned during network learning.

Layer-2: A node in this layer represents the antecedent part of a rule. Usually a T-norm operator is used in this node. The output of a layer 2 node represents the firing strength of the corresponding fuzzy rule.

Layer-3: Every node i in this layer is with a node function $\overline{w_i} f_i = \overline{w_i} (p_i x_1 + q_i x_2 + r_i)$ where $\overline{w_i}$ is the output of layer 2, and $\{p_i, q_i, r_i\}$ is the parameter set. A well-established way is to determine the consequent parameters is by using the least means squares algorithm.

Layer-4: The nodes in this layer compute the summation of all incoming signals $\sum_i w_i f_i$.

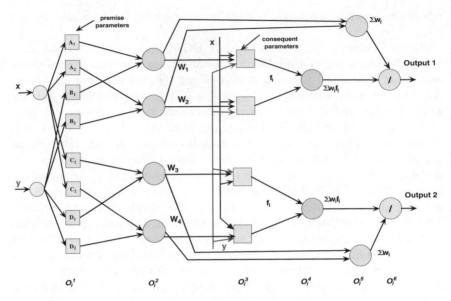

Fig. 5.1. Architecture of ANFIS with multiple outputs

Layer-5: The nodes in this layer compute the summation of all firing strength of the rule antecedents $\sum w_i$.

Layer-6: The nodes in this layer computes the overall output $(Output_i) = \frac{\sum_i w_i f_i}{\sum_i w_i}$.

ANFIS uses a mixture of backpropagation to learn the premise parameters and least mean square estimation to determine the consequent parameters. A step in the learning procedure has two parts: In the first part the input patterns are propagated, and the optimal conclusion parameters are estimated by an iterative least mean square procedure, while the antecedent parameters (membership functions) are assumed to be fixed for the current cycle through the training set. In the second part the patterns are propagated again, and in this epoch, backpropagation is used to modify the antecedent parameters, while the conclusion parameters remain fixed. This procedure is then iterated. Please refer to [11] for details of the learning algorithm.

5.3 Hard Computing

We used two popular hard computing techniques namely Multivariate Adaptive Regression Splines (MARS) and Classification and Regression Trees (CART). We also used a concurrent hybrid system involving MARS and CART.

5.3.1 Multivariate Adaptive Regression Splines (MARS)

MARS is a powerful well-established adaptive regression technique, which is known for its speed and accuracy [5, 8, 14]. The MARS model is a spline regression model that uses a specific class of basis functions as predictors in place of the original data. The MARS basis function transform makes it possible to selectively blank out certain regions of a variable by making them zero, allowing MARS to focus on specific sub-regions of the data. MARS excels at finding optimal variable transformations and interactions, as well as the complex data structure that often hides in high-dimensional data. A key concept underlying the spline is the knot. A knot marks the end of one region of data and the beginning of another. Thus, the knot is where the behavior of the function changes. Between knots, the model could be global (e.g., linear regression). In a classical spline, the knots are predetermined and evenly spaced, whereas in MARS, the knots are determined by a search procedure. Only as many knots as needed are included in a MARS model. If a straight line is a good fit, there will be no interior knots. In MARS, however, there is always at least one "pseudo" knot that corresponds to the smallest observed value of the predictor. Finding the one best knot in a simple regression is a straightforward search problem: simply examine a large number of potential knots and choose the one with the best R^2. However, finding the best pair of knots requires far more computation, and finding the best set of knots when the actual number needed is unknown is an even more challenging task [14].

MARS finds the location and number of needed knots in a forward/backward stepwise fashion. A model, which is clearly over fit with too many knots, is generated first, then, those knots that contribute least to the overall fit are removed. Thus, the forward knot selection will include many incorrect knot locations, but these erroneous knots will eventually, be deleted from the model in the backwards pruning step (although this is not guaranteed).

In MARS, Basis Functions (BFs) are the machinery used for generalizing the search for knots. BFs are a set of functions used to represent the information contained in one or more variables. Much like principal components, BFs essentially re-express the relationship of the predictor variables with the target variable. The hockey stick BF, the core building block of the MARS model is often applied to a single variable multiple times. The hockey stick function maps variable X^*: max $(0, X -c)$, or max $(0, c - X)$ where X^* is set to 0 for all values of X up to some threshold value c and X^* is equal to X for all values of X greater than c. (Actually X^* is equal to the amount by which X exceeds threshold c). The second form generates a mirror image of the first.

MARS generates basis functions by searching in a stepwise manner. It starts with just a constant in the model and then begins the search for a variable-knot combination that improves the model the most (or, alternatively, worsens the model the least). The improvement is measured in part by the change in Mean Squared Error (MSE). Adding a basis function always reduces the MSE. MARS searches for a pair of hockey stick basis functions, the primary and mirror image, even though only one might be linearly independent of the other terms. This search is then repeated, with MARS searching for the best variable to add given the basis functions already in the model. The brute search process theoretically continues until every possible basis function has been added to the model. In practice, the user specifies an upper limit for the number of knots to be generated in the forward stage. The limit should be large enough to ensure that the true model can be captured. A good rule of thumb for determining the minimum number is three to four times the number of basis functions in the optimal model. This limit may have to be set by trial and error.

5.3.2 Classification and Regression Trees (CART)

Tree-based models are useful for both classification and regression problems [6]. In these problems, there is a set of classification or predictor variables (X_i) and a dependent variable (Y). The X_i variables may be a mixture of nominal and/or ordinal scales (or code intervals of equal-interval scale) and Y a quantitative or a qualitative (i.e., nominal or categorical) variable.

The CART methodology is technically known as binary recursive partitioning [15]. The process is binary because parent nodes are always split into exactly two child nodes and recursive because the process can be repeated by treating each child node as a parent. The key elements of a CART analysis are a set of rules for:

- splitting each node in a tree;
- deciding when a tree is complete; and
- assigning each terminal node to a class outcome (or predicted value for regression)

CART is the most advanced decision-tree technology for data analysis, preprocessing and predictive modeling. CART is a robust data-analysis tool that automatically searches for important patterns and relationships and quickly uncovers hidden structure even in highly complex data. CART's binary decision trees are more sparing with data and detect more structure before further splitting is impossible or stopped. Splitting is impossible if only one case remains in a particular node or if all the cases in that node are exact copies of each other (on predictor variables). CART also allows splitting to be stopped for several other reasons, including that a node has too few cases.

Once a terminal node is found we must decide how to classify all cases falling within it. One simple criterion is the plurality rule: the group with the greatest representation determines the class assignment. CART goes a step further: because each node has the potential for being a terminal node, a class assignment is made for every node whether it is terminal or not. The rules of class assignment can be modified from simple plurality to account for the costs of making a mistake in classification and to adjust for over- or under-sampling from certain classes. A common technique among the first generation of tree classifiers was to continue splitting nodes (growing the tree) until some goodness-of-split criterion failed to be met. When the quality of a particular split fell below a certain threshold, the tree was not grown further along that branch. When all branches from the root reached terminal nodes, the tree was considered complete. Once a maximal tree is generated, it examines smaller trees obtained by pruning away branches of the maximal tree. Once the maximal tree is grown and a set of sub-trees is derived from it, CART determines the best tree by testing for error rates or costs. With sufficient data, the simplest method is to divide the sample into learning and test sub-samples. The learning sample is used to grow an overly large tree. The test sample is then used to estimate the rate at which cases are misclassified (possibly adjusted by misclassification costs). The misclassification error rate is calculated for the largest tree and also for every sub-tree. The best sub-tree is the one with the lowest or near-lowest cost, which may be a relatively small tree. Cross validation is used if data are insufficient for a separate test sample.

In the search for patterns in databases it is essential to avoid the trap of over fitting or finding patterns that apply only to the training data. CART's embedded test disciplines ensure that the patterns found will hold up when applied to new data. Further, the testing and selection of the optimal tree are an integral part of the CART algorithm. CART handles missing values in the database by substituting surrogate splitters, which are back-up rules that closely mimic the action of primary splitting rules. The surrogate splitter contains information that is typically similar to what would be found in the primary splitter.

5.3.3 Hybrid CART-MARS Model

CART and MARS could be integrated to work in a cooperative or concurrent environment. In a cooperative environment, CART plays an important role during the initialization of the prediction model. CART would go to the background after supplying some important variable information to MARS for building up the model. Thereafter MARS model works independently for further prediction. This sort of combination might be useful when not much variation is expected in the forex data. In a concurrent mode, CART and MARS are not independent. CART continuously provides intelligent variable information to improve the MARS prediction accuracy. This combination might be helpful when the forex data is continuously changing and requires constant updating of the prediction model.

We used the concurrent model where the forex values are fed to CART to provide some additional variable information to MARS. For modeling the forex data, we supplied MARS with the "node" information generated by CART. Figure 5.2 illustrates the hybrid CART-MARS model for predicting the forex values. As shown in Fig. 5.3, terminal nodes are numbered left to

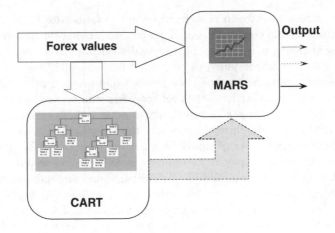

Fig. 5.2. Hybrid cooperative CART-MARS model for forex monitoring

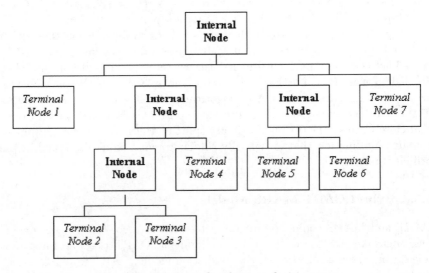

Fig. 5.3. Labeling of nodes in a decision tree

right starting with 1. All the data set records are assigned to one of the terminal nodes, which represent the particular class or subset. The training data together with this node information were supplied for training MARS.

5.4 Experiment Setup and Results

The data for our study were the monthly average forex rates from January 1981 to April 2001. We considered the exchange rates of the Australian dollar with respect to the Japanese yen, US Dollar, UK pound, Singapore dollar and

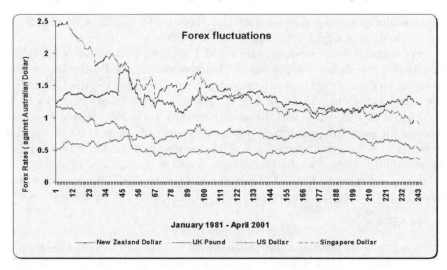

Fig. 5.4. Forex values from January 1981–April 2001 for four currencies

New Zealand dollar. Figure 5.4 shows the forex fluctuations during the period January 1981–April 2001 for the four different currencies. Due to scaling problems, Japanese Yen is not shown in Fig. 5.4. The experiment system consists of two stages: training the prediction systems and performance evaluation. For training the neural network, neuro-fuzzy model, MARS, CART and hybrid CART-MARS model, we selected the 'month, and 'previous month's forex rate' for the five currencies as input variables (total 6 input variables) and the 'current months forex rate' as output variable. We randomly extracted 70% of the data for training the prediction models and the remaining for testing purposes. The test data was then passed through the trained models to evaluate the prediction efficiency. Our objective is to develop an efficient and accurate forex prediction system capable of producing a reliable forecast. The required time-resolution of the forecast is monthly, and the required time-span of the forecast is one month ahead. This means that the system should be able to predict the forex rates one month ahead based on the values of the previous month.

5.4.1 Training the Different Computing Models

- **Soft computing models**

Our preliminary experiments helped us to formulate a feedforward neural network with 1 input layer, 2 hidden layers and an output layer [6-14-14-1]. The parameters of the neural network were decided after a trial and error approach. Input layer consists of 6 neurons corresponding to the input variables. The first and second hidden layers consist of 14 neurons respectively using tanh-sigmoidal activation functions. Training was terminated after 2000 epochs and we achieved a training error of 0.0251.

For training the neuro-fuzzy (NF) model, we used 4 Gaussian membership functions for each input variables and 16 rules were learned using the hybrid training method. Training was terminated after 30 epochs. For the NF model, we achieved training RMSE of 0.0248. The developed Takagi-Sugeno FIS is illustrated in Fig. 12. While the neural network took 200 seconds for 2000 epochs training, the neuro-fuzzy model took only 35 seconds for 30 epochs training. An important advantage of the neuro-fuzzy model is its easy interpretability using the 16 if-then rules which is graphically illustrated in Fig. 5.5. Referring to Fig. 5.5, the top section depicts the 6 input variables and the bottom section illustrates the 5 outputs (different currencies).

- **MARS**

We used 30 basis functions and to obtain the best possible prediction results (lowest RMSE), we sacrificed the speed (minimum completion time). It took almost 1 second to train the different forex prediction models.

- **CART**

We selected the minimum cost tree regardless the size of the tree. Figure 5.6 illustrates the variation of error as the numbers of nodes are increased. It took 3 seconds for developing the CART model for each prediction. For NZ dollar prediction, the developed tree has 7 terminal nodes as shown in Fig. 5.7.

- **Hybrid MARS-CART**

In the hybrid approach the data sets were first passed through CART and the node information were generated.The training data together with the node information (7 variables) were supplied for training MARS.

5.4.2 Test Results

Table 5.1 summarizes the performances of neural network, neuro-fuzzy model, MARS, CART and hybrid CART-MARS on the test set data. Figures 5.8, 5.9, 5.10, 5.11 and 5.12 illustrate the test results for forex prediction using MARS and CART and hybrid CART-MARS. The actual predicted values by each technique is plotted against the desired value for each currency.

5.5 Conclusions

In this chapter, we have investigated the performance of neural network, neuro-fuzzy system, MARS, CART and a hybrid CART-MARS technique for predicting the monthly average forex rates of US dollar, UK pounds, Singapore

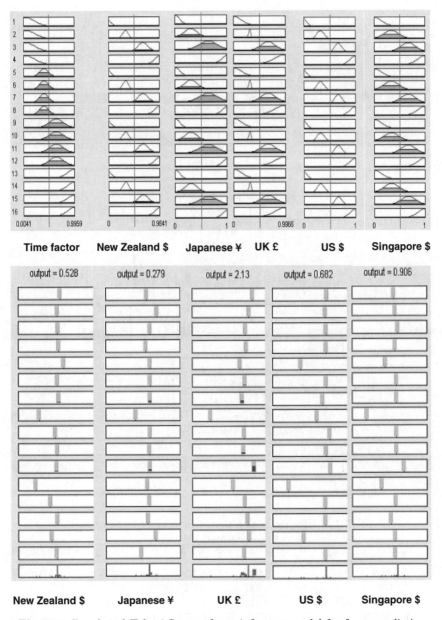

Fig. 5.5. Developed Takagi-Sugeno fuzzy inference model for forex prediction

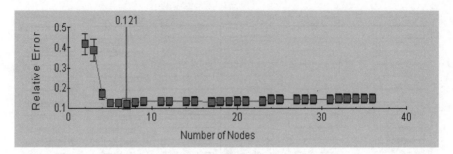

Fig. 5.6. Change in relative error when the number of nodes is increased

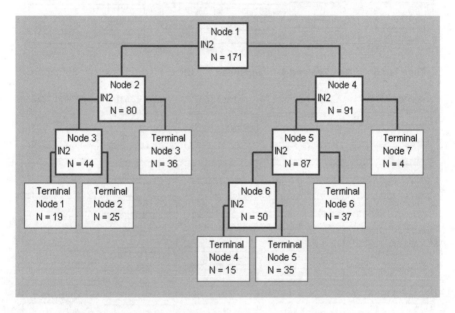

Fig. 5.7. Developed regression tree for NZ dollar prediction

Table 5.1. Performance of different models for forex prediction

System	Japan Yen	US $	UK £	Singapore $	NZ $
MARS	0.023	0.039	0.048	0.028	0.049
CART	0.037	0.037	0.063	0.033	0.041
Hybrid MARS-CART	0.016	0.027	0.035	0.026	0.035
ANN	0.028	0.034	0.023	0.030	0.021
Neuro-Fuzzy	0.026	0.034	0.037	0.029	0.020

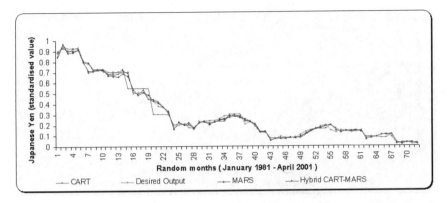

Fig. 5.8. Test results for Japanese Yen

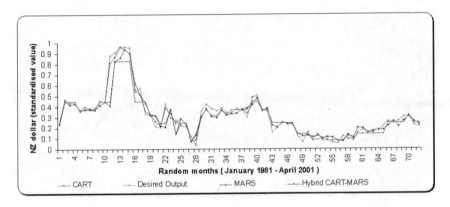

Fig. 5.9. Test results for New Zealand dollar

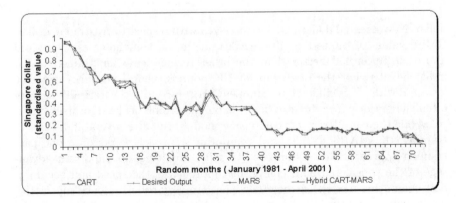

Fig. 5.10. Test results for Singapore dollar

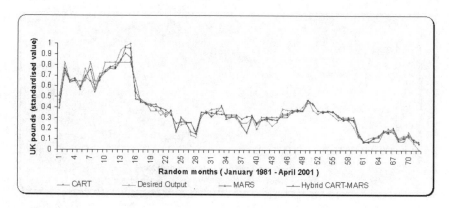

Fig. 5.11. Test results for UK pounds

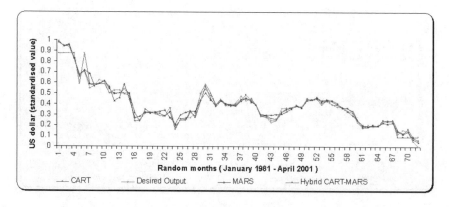

Fig. 5.12. Test results for US dollar

dollar, New Zealand dollar and Japanese yen with respect to Australian dollar. RMSE values of the test results reveal that, in most of the cases the hybrid approach performed better than the other techniques when trained/tested independently. For the prediction of UK pounds, neural networks gave the lowest RMSE. It is difficult to comment on the results theoretically as very often the performance directly depends on the profile of the data itself.

While the considered soft computing models, requires several iterations of training, the MARS/CART hard computing approach works on a one pass training approach. Hence compared to soft computing, an important advantage of the considered hard computing approach is the speed and accuracy. Soft computing models on the other hand are very robust. Hence superiority of the soft computing models will be on the robustness in particularly the easy interpretability (*if-then* rules) of the neuro-fuzzy models.

Our future research is targeted to study an ensemble approach by combining the outputs of different hybrid approaches.

5.6 Acknowledgements

The author wishes to thank the anonymous reviewers for their constructive comments which helped to improve the presentation of the chapter.

References

1. A. Abraham, Neuro-Fuzzy Systems: State-of-the-Art Modeling Techniques, Connectionist Models of Neurons, Learning Processes, and Artificial Intelligence, Lecture Notes in Computer Science, Jose Mira and Alberto Prieto (Eds.), Germany, Springer-Verlag, LNCS 2084, pp. 269–276, 2001.
2. A. Abraham, Analysis of Hybrid Soft and Hard Computing Techniques for Forex Monitoring Systems, 2002 IEEE International Conference on Fuzzy Systems, IEEE Press, pp. 1616–1622, 2002.
3. A. Abraham, M. Chowdury and S. Petrovic-Lazerevic, Australian Forex Market Analysis Using Connectionist Models, International Journal of Management, Vol. 29, pp. 18–22, 2003.
4. A. Abraham and M. Chowdhury, An Intelligent Forex Monitoring System, In Proceedings of IEEE International Conference on Info-tech and Info-net, Beijing, China, IEEE Press, pp. 523–528, 2001.
5. A. Abraham and D. Steinberg, MARS: Still an Alien Planet in Soft Computing?, Computational Science, Springer-Verlag Germany, Vassil N Alexandrov et al. (Editors), San Francisco, USA, pp. 235–244, 2001.
6. L. Breiman, J. Friedman, R. Olshen and C.J. Stone, Classification and Regression Trees, Chapman and Hall, New York, 1984.
7. A.V. Dormale, The Power of Money, Macmillan Press, London, 1997.
8. J.H. Friedman, Multivariate Adaptive Regression Splines, Annals of Statistics, Vol. 19, pp. 1–141, 1991.
9. J.O. Grabbe, International Financial Markets, Englewood Hills, Prentice Hall Inc., USA, 1996.
10. Introduction to Forex Market (2004), <http://www.forexcapital.com>, accessed on 22 March 2004.
11. S.R. Jang, C.T. Sun and E. Mizutani, Neuro-Fuzzy and Soft Computing: A Computational Approach to Learning and Machine Intelligence, Prentice Hall Inc., USA, 1997.
12. K. Longl and K. Walter, Electronic Currency Trading for Maximum Profit, Prima Money, Roseville, California, 2001.
13. A.F. Moller, A Scaled Conjugate Gradient Algorithm for Fast Supervised Learning, Neural Networks. 6: pp. 525–533, 1993.
14. D. Steinberg and P.L. Colla, MARS User Guide, San Diego, USA, Salford Systems Inc., 1999.
15. D. Steinberg and P.L. Colla, CART: Tree-Structured Nonparametric Data Analysis, San Diego, Salford Systems Inc., USA, 1995.
16. USFXM (2004): Foreign Exchange Market in the United States, http://www.newyorkfed.org/fxc/, Accessed on 22 March 2004.
17. L.A. Zadeh, Roles of Soft Computing and Fuzzy Logic in the Conception, Design and Deployment of Information/Intelligent Systems, Computational Intelligence: Soft Computing and Fuzzy-Neuro Integration with Applications, O. Kaynak, L.A. Zadeh, B. Turksen, I.J. Rudas (Eds.), pp. 1–9, 1998.

6

On the Stability and Sensitivity Analysis of Fuzzy Control Systems for Servo-Systems

R.-E. Precup[1] and S. Preitl[2]

[1] Department of Automation and Industrial Informatics,
   Faculty of Automation and Computers, "Politehnica" University of Timisoara,
   Bd. V. Parvan 2, 300223 Timisoara, Romania
   rprecup@aut.utt.ro, www.aut.utt.ro/~rprecup
[2] Department of Automation and Industrial Informatics,
   Faculty of Automation and Computers, "Politehnica" University of Timisoara,
   Bd. V. Parvan 2, 300223 Timisoara, Romania
   spreitl@aut.utt.ro, www.aut.utt.ro/~spreitl

Control systems must ensure in real-world applications good steady-state and dynamic performance. This is the reason why they need high quality servo-systems to perform the tasks of stabilization and tracking, and to deal with the problems created by the parameter variance or the nonlinearities. A way to fulfill these tasks is to employ fuzzy control. The development of fuzzy control systems is usually performed by heuristic means, incorporating human skills, but the drawback is in the lack of general-purpose development methods. A major problem, which follows from this way of developing fuzzy controllers, is the analysis of the structural properties of the control system, such as stability, controllability and robustness. This is the reason for the first aim of the chapter, to present stability analysis methods dedicated to fuzzy control systems for servo-systems: the state-space approach, the use of Popov's hyperstability theory, the circle criterion and the harmonic balance method. The second aim of the chapter is to perform the sensitivity analysis of fuzzy control systems with respect to the parametric variations of the controlled plant for a class of servo-systems based on the construction of sensitivity models; both of them, the stability and sensitivity analysis, provide useful information to the development of fuzzy control systems. The presented case studies concerning fuzzy controlled servo-systems validate the presented methods. Since some of the case studies deal with several fuzzy controllers, there are derived useful development conditions for these ones.

6.1 Introduction

The development of fuzzy control systems is usually performed by heuristic means, due to the lack of general development methods applicable to large categories of systems. A major problem, which follows from the heuristic method of developing fuzzy controllers, is the analysis of the structural properties of the control systems including the stability analysis and the sensitivity analysis with respect to the parametric variations of the controlled plant.

In the case of servo-systems the analysis of these properties becomes more important due to the very good steady-state and dynamic performance they must ensure. Therefore, the chapter aims a twofold goal. Firstly, as a general goal, it presents stability analysis methods dedicated to fuzzy control systems applied to servo-systems: the state-space approach, the use of Popov's hyperstability theory, the circle criterion and the harmonic balance method. Secondly, as a particular goal, the chapter performs the sensitivity analysis of fuzzy control systems with respect to the parametric variations of the controlled plant (CP) for a class of servo-systems based on the construction of sensitivity models. The considered fuzzy control systems contain fuzzy controllers, which are type-II fuzzy systems [1, 2], or Mamdani fuzzy controllers with singleton consequents.

The stability analysis of a fuzzy control system (abbreviated FCS) is justified because only a stable FCS can ensure the functionality of the plant and, furthermore, the disturbance reduction, guarantee desired steady states, and reduce the risk of implementing the fuzzy controller (FC).

Several approaches and good overviews [2–4] have been used for the stability analysis of fuzzy control systems, mainly based on the classical theory of nonlinear dynamic systems. These approaches include:

- the state-space approach, based on a linearized model of the nonlinear system [5–7];
- Popov's hyperstability theory [8, 9];
- Lyapunov's stability theory [2, 4];
- the circle criterion [3, 4, 8];
- the harmonic balance method [4, 10].

The sensitivity analysis of the FCSs with respect to the parametric variations of the CP is necessary because the behavior of these systems is generally reported as "robust" or "insensitive" without offering systematic analysis tools. The sensitivity analysis performed in the chapter is based on the idea of approximate equivalence between the FCSs and the linear control systems, in certain conditions. This is fully justified because of two reasons.

The first reason is related with the controller part of the FCS, where the approximate equivalence between linear and fuzzy controllers is generally acknowledged and widely accepted and used [11–13].

The second reason is related with the plant part of the FCS. The support for using an FC developed to control a plant having a linear or linearized model is in the fact that this plant model can be considered as a simplified model of a relatively complex model of the CP (for example, the servo-system as CP), having nonlinearities or variable parameters or being placed at the lower level of large-scale systems. The idea is in the fact that the plant is nonlinear but linearized in the vicinity of a set of operating points or of a trajectory. The plant model could be also uncertain or not well defined. The FC, as essentially nonlinear element, can compensate – based on the designers' experience – the model uncertainties, nonlinearities and parametric variations of the CP; it must not be seen as a goal in itself, but sometimes the only way to initially approach the control of such relatively complex plants (in particular, the servo-systems).

This chapter addresses the following topics. It will be treated in the following Sections the stability analysis methods based on the state-space approach, the use of Popov's hyperstability theory, the circle criterion and the harmonic balance method. The exemplification of the methods is done by case studies regarding FCs to control an electro-hydraulic servo-system. Then, in Sect. 6.6 an approach to the sensitivity analysis of the considered class of FCSs with respect to the parametric variations of the CP is presented. This approach is illustrated by a case study regarding the fuzzy control of a servo-system corresponding to the simplified dynamics used in mobile robots. The conclusions are drawn in Sect. 6.7.

6.2 A State-Space Approach to Stability Analysis

The presented stability analysis method (SAM) employing the state-state approach is based on a linearized approximate mathematical model of the nonlinear control system in the vicinity of its equilibrium points. The stability conditions are expressed by two indices that indicate the character of the equilibrium point (EP). For the sake of ensuring the global stability of the system, another index verifies the uniqueness of the EP. Among the advantages of this method are its simplicity, and its applicability to a large class of nonlinear systems. It can be easily applied to Single-Input Single-Output (SISO) systems, up to second-order, or higher-order systems that can be reduced to second-order ones.

Without the loss of generality, the presentation in the sequel will be dedicated to single input second-order plants. This is the reason why the geometric interpretations taken from the qualitative theory of nonlinear dynamic systems become easily accessible. The application of the method to a third-order system is presented in [7].

Consider the dynamics of the CP can be described by the following state equation:

$$\dot{\mathbf{x}} = \mathbf{f}(\mathbf{x}) + \mathbf{b}u , \qquad (6.1)$$

where: $\mathbf{x} = [\mathbf{x}_1, \mathbf{x}_2, \ldots, \mathbf{x}_n]^{\mathrm{T}} \in \Re^n$ – the state vector; $\mathbf{b} = [\mathbf{b}_1, \mathbf{b}_2, \ldots, \mathbf{b}_n]^{\mathrm{T}}$ – an $[n, 1]$-dimensional vector of constant coefficients; $u = \varphi(\mathbf{i})$ or $u = \phi(\mathbf{x})$ – the control signal expressed by means of the nonlinear functions φ or ϕ, with $\varphi, \phi : \Re^n \to \Re$; i – the input vector in the form $\mathbf{i} = \mathbf{T} \cdot [\mathbf{r}; \mathbf{x}^T]^T$, with $\mathbf{T}$ – an $[n, n+1]$-dimensional constant matrix (not necessarily to be expressed); r – the reference input; $\mathbf{f} : \Re^n \to \Re^n, \mathbf{f}(\mathbf{x}) = [\mathbf{f_1}(\mathbf{x}), \mathbf{f_2}(\mathbf{x}), \dots, \mathbf{f_n}(\mathbf{x})]^T$ – the CP function; the only constraint imposed on f is that it must be continuous and partially differentiable [5], but it can be accepted in certain conditions to be piecewise linear.

For the sake of simplicity the time variable (t) has been omitted in (6.1), and it will be omitted in other equations in this chapter.

All relations in this Section will be particularized as follows for second-order systems $(n = 2)$.

The EPs are found as part of the solutions of (6.2):

$$\frac{d\mathbf{x}}{dt} = \mathbf{0} . \tag{6.2}$$

It can be assumed that the origin of the state space is among these solutions, which is equivalent to:

$$\mathbf{f}(\mathbf{0}) = [\mathbf{f_1}(\mathbf{0}, \mathbf{f_2}(\mathbf{0})]^T = \mathbf{0}^T = [0, 0]^T , \tag{6.3}$$

where f_1 and f_1 are the two components of $\mathbf{f}$.

This relation is not a constraint because any solution can be translated to the origin by a coordinate transformation.

The stability conditions for the origin EP are derived from the stability theorem, applied to a linearized approximation of the control system function. So, based on a Lyapunov criterion, if the EP of the linearized system is as-ymptotically stable, then it will also be asymptotically stable for the original nonlinear system [5].

The Jacobian matrix $\mathbf{J}(\mathbf{x_0})$ of the control system can be expressed in terms of (6.4):

$$\mathbf{J}(\mathbf{x_0}) = \begin{bmatrix} \frac{\partial F_1}{\partial x_1} & \frac{\partial F_1}{\partial x_2} \\ \frac{\partial F_2}{\partial x_1} & \frac{\partial F_2}{\partial x_2} \end{bmatrix}_{\mathbf{x_0}} , \tag{6.4}$$

where $\mathbf{F}$ corresponds to the right-hand term of (6.1):

$$\mathbf{F}(\mathbf{x}) = \mathbf{f}(\mathbf{x}) + \mathbf{bu} = [\mathbf{F_1}(\mathbf{x}), \mathbf{F_2}(\mathbf{x})]^T . \tag{6.5}$$

The characteristic polynomial of $\mathbf{J}(\mathbf{x_0})$ is $\mu(s)$:

$$\mu(s) = det(s\mathbf{I} - \mathbf{J}(\mathbf{x_0})) = s^2 - \mathbf{tr}(\mathbf{J}(\mathbf{x_0}))s + \mathbf{det}(\mathbf{J}(\mathbf{x_0})) , \tag{6.6}$$

where:

$$tr(\mathbf{J}(\mathbf{x_0})) = \mathbf{a_{11}} + \mathbf{a_{22}}, \mathbf{det}(\mathbf{J}(\mathbf{x_0})) = \mathbf{a_{11}a_{22}} - \mathbf{a_{12}a_{21}} , \tag{6.7}$$

a_{11}, a_{12}, a_{21} and a_{22} are the elements of $\mathbf{J}(\mathbf{x_0})$ calculated in $\mathbf{x_0} = \mathbf{0}$.

The conditions for $\mathbf{J}(\mathbf{x_0})$ to have negative real part eigenvalues (or, in other words, for $\mu(s)$ to be a Hurwitz polynomial) are equivalent to (6.8):

$$I_1 = -tr(\mathbf{J}(\mathbf{x_0})) = -(\mathbf{a_{11}} + \mathbf{a_{22}}), I'_1 = \det(\mathbf{J}(\mathbf{x_0})) = \mathbf{a_{11}a_{22}} - \mathbf{a_{12}a_{21}} . \quad (6.8)$$

The stability of the EP placed in the origin is ensured if and only if both indices $\{I_1, I'_1\}$ have strictly positive values.

It is well known that the closed-loop system will be globally stable if the origin remains the only stable EP. It can be shown by a geometrical approach that another solution of (6.2) exists only when the controller vector field $\mathbf{b}\ \varphi(\mathbf{i})$ compensates for the plant component, i.e. the vector field $\mathbf{f}(\mathbf{x})$ [5]. This is possible only in regions when both vectors have the same direction. The subspace A of the state space for which this condition is fulfilled is defined according to (6.9):

$$A = \{\mathbf{x} \in \Re^2 | \mathbf{f_1}(\mathbf{x})\mathbf{b_2} = \mathbf{f_2}(\mathbf{x})\mathbf{b_1}\} , \quad (6.9)$$

with b_1 and b_2 being the components of $\mathbf{b}$. So, a third stability index is defined on the subspace A:

$$I_2 = \min\{\|\mathbf{f}(\mathbf{x}) + \mathbf{bu}\| | \mathbf{x} \in A \setminus \mathbf{SR}\}, \mathbf{SR} = \{\mathbf{x} \in \Re^2 | \|\mathbf{f}(\mathbf{x}) + \mathbf{bu}\| = \mathbf{0}\} , \quad (6.10)$$

where SR – a small sub-region around the origin, and $\|\mathbf{x}\|$ – the Euclidean norm of $\mathbf{x}$.

The stability conditions are expressed in terms of the following inequalities:

$$I_1 > 0, I'_1 > 0, I_2 > 0 . \quad (6.11)$$

The magnitudes of all three indices $\{I_1, I'_1, I_2\}$ indicate the stability margin of the control system.

Beside the above presented stability indices, a fourth stability index, that measures the relative degree of stability for second-order systems, has been presented in [7]. The indices I_1 and I'_1 ensure only the position of closed-loop system eigenvalues (poles) p^*_1 and p^*_2 in the negative real part half-plane of the complex plane. However, it is known that the distance from the complex eigenvalues to the imaginary axis is related to the damping and overshoot of the system. The latter might be too large to be considered acceptable when this distance is too small [14].

By computing the tangent of the angle θ (Fig. 6.1) and by imposing a maximum value of this angle, one can verify that the overshoot and damping of the control system are within certain limits. There exists a proportional but not linear dependence between the values of θ ($\tan(\theta)$) and the overshoot, and the maximum admissible overshoot/angle depends on the system involved.

The fourth index can be defined according to (6.12):

$$I''_1 = \tan(\theta) = \frac{\sqrt{4I'_1 - I_1^2}}{I_1}, I_1^2 < 4I'_1 . \quad (6.12)$$

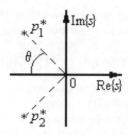

Fig. 6.1. Interpretation of eigenvalue positions for second-order systems

Unlike the other three indices, a more stable character of the control system is obtained for smaller values of I_1''.

From the constraint (6.13):

$$I_1'' < \tan(\theta_{max}) \tag{6.13}$$

and from the definition (6.12), the inequalities (6.14) will result:

$$4I_1' \cos^2(\theta_{max}) \le I_1^2 \le 4I_1' . \tag{6.14}$$

By expressing I_1 and I_1' in terms of the parameters of the input and output membership functions of the FC, the relation (6.14) represents a good development condition.

Summarizing all aspects presented before, the SAM consists of the following steps:

step 1: express the mathematical model of the CP in the form (6.1);
step 2: compute the values of the function $\mathbf{F}(\mathbf{x})$ in (6.5) as function of the controller structure, and the Jacobian matrix $\mathbf{J}(\mathbf{x_0})$ of the closed-loop system;
step 3: compute the values of the stability indices $\{I_1, I_1'\}$ in (6.8);
step 4: set the values of the subspace A and of its sub-region SR, and compute the value of the stability index I_2 in (6.10);
step 5: the FCS is stable if the conditions (6.11) are fulfilled.

It must be highlighted that the method offers sufficient stability conditions, and the steps 2 and 3 are the most computationally expensive ones.

To apply the SAM it is considered the controlled plant part of an electro-hydraulic servo-system (EHS) with the informational block diagram presented in Fig. 6.2.

The significance of blocks and variables in Fig. 6.2 is: NL 1 $\cdots$ NL 5 – the nonlinearities, realized technological and measured experimental [15]; EHS – the electro-hydraulic converter; SVD – the slide-valve distributor; MSM – the main servo-motor; M 1 and M 2 – the measuring devices; u – the control signal, y – the controlled output; x_1 and x_2 – the state variables; x_{1M} and x_{2M} – the measured state variables. The CP parameters are [16]: $u_1 = 10V$,

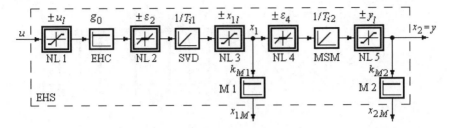

Fig. 6.2. Structure of electro-hydraulic servo-system

$g_0 = 0.0625\frac{mm}{V}$, $\varepsilon_2 = 0.02\,mm$, $\varepsilon_4 = 0.2\,mm$, $x_{1l} = 21.8\,mm$, $y_l = 210\,mm$, $T_{i1} = 0.001872\,sec$, $T_{i2} = 0.0756\,sec$, $k_{M1} = 0.2\frac{V}{mm}$, $k_{M2} = 0.032\frac{V}{mm}$.

Due to the very large values of the insensitivity and of the large linear domains of NL 1, NL 3 and NL 5 (in the conditions of small variations of the variables) by omitting the nonlinearities in Fig. 6.2, the linear state mathematical model (MM) of the CP (the step 1 of the SAM) results as:

$$\begin{bmatrix} \dot{x}_1 \\ \dot{x}_2 \end{bmatrix} + \begin{bmatrix} 0 & 0 \\ a & 0 \end{bmatrix} \cdot \begin{bmatrix} x_1 \\ x_2 \end{bmatrix} + \begin{bmatrix} b^* \\ 0 \end{bmatrix} \cdot u , \tag{6.15}$$

$$y = x_2, \quad a = \frac{1}{T_{i2}} = 14.05, \quad b^* = \frac{g_0}{T_{i1}} = 26.04 .$$

The structure of the FCS is illustrated in Fig. 6.3, where: r – the reference input; e – the control error, i_1 and i_2 – the FC inputs, $\mathbf{i} = [\mathbf{i_1}, \mathbf{i_2}]^{\mathbf{T}}$. The block FC in Fig. 6.3 is a state feedback fuzzy controller that can be developed by starting with the development of a conventional state feedback controller to stabilize the EHS, and by applying the modal equivalence principle [12].

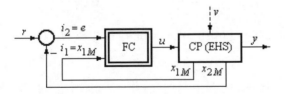

Fig. 6.3. Structure of fuzzy control system

The FC is characterized by the input and output membership functions with the shapes presented in Fig. 6.4, it employs Mamdani's MAX-MIN compositional rule of inference assisted by the rule base presented in Table 6.1 as decision table, and the center of gravity method for defuzzification.

In the step 2 of the SAM it is expressed $\mathbf{F}(\mathbf{x})$:

$$\mathbf{F}(\mathbf{x}) = [\mathbf{b}^* \phi(\mathbf{x_1}, \mathbf{x_2}), \mathbf{a} \cdot \mathbf{x_1}]^{\mathbf{T}} , \tag{6.16}$$

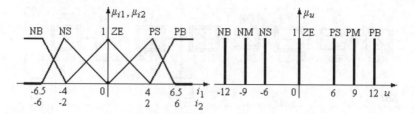

Fig. 6.4. Input and output membership functions of FC in Fig. 6.3

Table 6.1. Decision table of FC in Fig. 6.3

$i_1 \backslash i_2$	NB	NS	ZE	PS	PB
PB	NB	NM	NM	NS	PS
PS	NM	NS	NS	ZE	PB
ZE	NM	NS	ZE	PS	PM
NS	NB	ZE	PS	PS	PM
NB	NS	PS	PM	PM	PB

and the Jacobian matrix $\mathbf{J}(\mathbf{x_0})$ is obtained by applying (6.4) in its particular form:

$$\mathbf{J}(\mathbf{x_0}) = \begin{bmatrix} b^* \frac{\partial \phi}{\partial x_1} & b^* \frac{\partial \phi}{x_2} \\ a & 0 \end{bmatrix}_{\mathbf{x_0}}, \quad \mathbf{x_0} = \mathbf{0} . \tag{6.17}$$

By using the connections between the state variables and the FC inputs (Fig. 6.2 and Fig. 6.3), the connections (6.18) can be derived (with partial derivatives in origin):

$$\frac{\partial \phi}{\partial x_1} = \frac{\partial \varphi}{\partial i_1}, \frac{\partial \phi}{\partial x_2} = -k_{M2} \frac{\partial \varphi}{\partial i_2} . \tag{6.18}$$

By using (6.17) and (6.18), the stability indices $\{I_1, I_1'\}$ (in the step 3) can be obtained:

$$I_1 = -a \cdot b^* \left(\frac{\partial \varphi}{\partial i_1} \right)_{\mathbf{x_0}}, I_1' = b^* k_{M2} \left(\frac{\partial \varphi}{\partial i_2} \right)_{\mathbf{x_0}}, \tag{6.19}$$

but this requires the calculation of the partial derivatives $(\frac{\partial \varphi}{\partial i_1})_{\mathbf{x_0}}$ and $(\frac{\partial \varphi}{\partial i_2})_{\mathbf{x_0}}$.

By using (6.19), the first two stability conditions in (6.11) are equivalent to:

$$\left(\frac{\partial \varphi}{\partial i_1} \right)_{\mathbf{x_0}} < 0, \left(\frac{\partial \varphi}{\partial i_2} \right)_{\mathbf{x_0}} > 0 . \tag{6.20}$$

The calculation of the two partial derivatives in (6.20) can be performed for the activated rules in the vicinity of the origin EP by using the accepted inference and defuzzification methods. The division of the input space in several subspaces accompanies this. In all situations the stability conditions (6.20) are fulfilled.

In the step 4 of the SAM the subspace A and the stability index I_2 become:

$$A = \{\mathbf{x} = [\mathbf{x_1}, \mathbf{x_2}]^{\mathbf{T}} \in \Re^2 | \mathbf{x_1} = \mathbf{0}\},$$
$$I_2 = \min\{|\varphi(0, x_2)| \,|\mathbf{x} \in A \setminus SR\}. \tag{6.21}$$

Setting the value of SR and the computation of I_2 represent relatively difficult tasks in this case. Digital simulations can assist these tasks.

Digital simulations performed for the significant regimes of the FCS prove that the third stability condition in (6.11) is fulfilled.

By concluding the presented stability analysis points of view (in the step 5 of the SAM), the FCS considered in this case study is stable.

6.3 Stability Analysis Method Based on Popov's Hyperstablity Theory

The SAM based on Popov's hyperstability theory is applied to FCSs to control SISO plants when employing PI-fuzzy controllers (PI-FCs). The structure of the considered FCS is a conventional one, presented in Fig. 6.5 (a), where: r – the reference input, y – the controlled output, e – the control error, u – the control signal, d_1, d_2 and d_3 – the disturbance inputs.

The controlled plant includes the actuator and the measuring devices. The application of an FC, when conditions for linear operating regimes of the plant are validated, determines the FCS to be considered as a Lure-Postnikov type nonlinear control system (see the example in [17]).

The PI-FC represents a discrete-time FC with dynamics, introduced by the numerical differentiation of the control error e_k expressed as the increment of control error, $\triangle e_k = e_k - e_{k-1}$, and by the numerical integration of the increment of control signal, $\triangle u_k$. The structure of the considered PI-FC is illustrated in Fig. 6.5 (b), where B-FC represents the basic fuzzy controller, without dynamics.

The block B-FC is a nonlinear two inputs-single output (TISO) system, which includes among its nonlinearities the scaling of inputs and output as part of its fuzzification module. The fuzzification is solved in terms of the regularly distributed (here) input and output membership functions illustrated

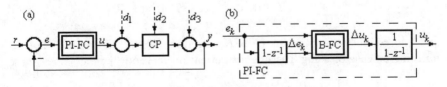

Fig. 6.5. Structure of FCS and of PI-FC

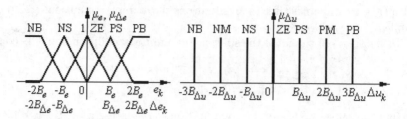

Fig. 6.6. Membership functions of B-FC input and output linguistic variables

in Fig. 6.6. Other distributions of the membership functions can modify in a desired way the controller nonlinearities.

The inference engine in B-FC employs Mamdani's MAX-MIN compositional rule of inference assisted by the rule base presented in Table 6.2, and the center of gravity method for singletons is used for defuzzification.

Table 6.2. Decision table of B-FC

$i_1 \backslash i_2$	NB	NS	ZE	PS	PB
PB	NB	NM	NM	NS	PS
PS	NM	NS	NS	ZE	PB
ZE	NM	NS	ZE	PS	PM
NS	NB	ZE	PS	PS	PM
NB	NS	PS	PM	PM	PB

The start of the development of the PI-FC is in the expression of the discrete-time equation of a digital PI controller in its incremental version:

$$\triangle u_k = K_P \triangle e_k + K_I e_k = K_P(\triangle e_k + \alpha \cdot e_k) , \qquad (6.22)$$

where k is the index of the current sampling interval.

In the case of a quasi-continuous digital PI controller the parameters K_P, K_I and α can be computed as functions of the parameters k_C (gain) and T_i (integral time constant) of a basic original continuous-time PI controller having the transfer function $H_C(s)$:

$$H_C(s) = \frac{k_C}{sT_i}(1 + sT_i) , \qquad (6.23)$$

and the connections between $\{K_P, K_I, \alpha\}$ and $\{k_C, T_i\}$ have the following form in the case of using Tustin's discretization method.

$$K_P = k_C \left[1 - \frac{T_s}{2T_i}\right], K_I = k_C \frac{T_s}{T_i}, \alpha = \frac{K_I}{K_P} = \frac{2T_s}{2T_i - T_s} , \qquad (6.24)$$

with T_s – the sampling period chosen in accordance with the requirements of quasi-continuous digital control [18].

The design relations for the PI-FC are obtained by the application of the modal equivalence principle [12] particularized to:

$$B_{\triangle e} = \alpha B_e, B_{\triangle u} = K_I B_e ,\qquad (6.25)$$

where the free parameter B_e represents the designers' option. Using the experience in controlling the plant one can choose the value of this parameter, but firstly it must be chosen to ensure the aim of a stable FCS.

The CP is supposed to be characterized by the n-th order discrete-time SISO linear time-invariant state MM (6.26) including the zero-order hold:

$$\mathbf{x_{k+1}} = \mathbf{A} \cdot \mathbf{x_k} + \mathbf{b} \cdot \mathbf{u_k} ,$$
$$y_k = \mathbf{c^T} \cdot \mathbf{x_k} ,\qquad (6.26)$$

where: u_k – the control signal; y_k – the controlled output; $\mathbf{x_k}$ – the state vector, $\dim \mathbf{x_k} = (\mathbf{n,1})$; $\mathbf{A}$, $\mathbf{b}$, $\mathbf{c^T}$ – matrices with the dimensions: $\dim \mathbf{A} = (\mathbf{n,n})$, $\dim \mathbf{b} = (\mathbf{n,1})$, $\dim \mathbf{c^T} = (\mathbf{1,n})$.

To derive the SAM it is necessary to transform the initial FCS structure into a multivariable one because the block B-FC in Fig. 6.5 is a TISO system. This modified FCS structure is illustrated in Fig. 6.7 (a), where the dynamics of the fuzzy controller (its linearized part) is transferred to the plant (CP) resulting in the extended controlled plant (ECP, a linear one).

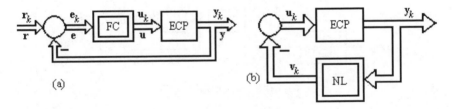

Fig. 6.7. FCS modified structure (**a**); stability analysis-oriented structure (**b**)

The vectors in Fig. 6.7 (a) represent: $\mathbf{r_k}$ – the reference input vector, $\mathbf{e_k}$ – the control error vector, $\mathbf{y_k}$ – the controlled output vector, $\mathbf{u_k}$ – the control signal vector.

For the general use (in the continuous time, too) the index k may be omitted.

These vectors are defined as follows:

$$\mathbf{r_k} = [\mathbf{r_k}, \triangle \mathbf{r_k}]^T, \mathbf{e_k} = [\mathbf{e_k}, \triangle \mathbf{e_k}]^T, \mathbf{y_k} = [\mathbf{y_k}, \triangle \mathbf{y_k}]^T ,\qquad (6.27)$$

where $\triangle m_k = m_k - m_{k-1}$ stands for the increment of the variable m_k.

In relation with Fig. 6.7 (a), the block FC is characterized by the nonlinear input-output static map $\mathbf{F}$:

$$\mathbf{F} : \Re^2 \to \Re^2, \mathbf{F}(\mathbf{e_k}) = [\mathbf{f}(\mathbf{e_k}), \mathbf{0}]^{\mathbf{T}} , \qquad (6.28)$$

where f ($f : \Re^2 \to \Re$) is the input-output static map of the nonlinear TISO system B-FC in Fig. 6.5.

As it can be seen in (6.27), all variables in the FCS structure (presented in Fig. 6.7 (a)) have two components. This requires the introduction of a fictitious control signal, supplementary to the outputs of the block B-FC, for obtaining an equal number of inputs and outputs as required by the hyperstability theory in the multivariable case.

Generally, the structure involved in the stability analysis of an unforced nonlinear control system ($\mathbf{r_k} = \mathbf{0}$ and the disturbance inputs are also zero) is presented in Fig. 6.7 (b). The block NL in Fig. 6.7 (b) represents a static nonlinearity due to the nonlinear part without dynamics of the block FC in Fig. 6.7 (a). The connections between the variables of the control system structures in Fig. 6.7 are:

$$\mathbf{v_k} = -\mathbf{u_k} = -\mathbf{F}(\mathbf{e_k}), \mathbf{y_k} = -\mathbf{e_k} , \qquad (6.29)$$

with the second component in $\mathbf{F}$ being always zero to neglect the effect of the fictitious control signal.

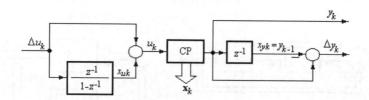

Fig. 6.8. Structure of ECP

The MM of the ECP can be derived by firstly defining the additional state variables $\{x_{uk}, x_{yk}\}$ according to Fig. 6.8. Then, the extended state vector $\mathbf{x_k^E}$ and the control signal vector $\mathbf{u_k^E}$ can be expressed in terms of:

$$\mathbf{x_k^E} = \left[\mathbf{x_k^T}, \mathbf{x_{uk}}, \mathbf{x_{yk}}\right]^{\mathbf{T}}, \mathbf{u_k^E} = [\triangle\mathbf{u_k}, \triangle\mathbf{u_{fk}}]^{\mathbf{T}} , \qquad (6.30)$$

where $\triangle u_{fk}$ stands for the fictitious increment of control signal.

By using the structure presented in Fig. 6.8, the $n + 2$-th order discrete-time state MM of the ECP becomes (6.31):

$$\mathbf{x_{k+1}^E} = \mathbf{A^E}\mathbf{x_k^E} + \mathbf{B^E}\mathbf{u_k^E} , \qquad (6.31)$$

$$\mathbf{y_k^E} = \mathbf{C^E}\mathbf{x_k^E} ,$$

with the matrices $\mathbf{A}^{\mathbf{E}}$ (dim $\mathbf{A}^{\mathbf{E}} = (\mathbf{n}+2, \mathbf{n}+2)$), $\mathbf{B}^{\mathbf{E}}$ (dim $\mathbf{B}^{\mathbf{E}} = (\mathbf{n}+2, 2)$) and $\mathbf{C}^{\mathbf{E}}$ (dim $\mathbf{C}^{\mathbf{E}} = (2, \mathbf{n}+2)$) as follows:

$$
\mathbf{A}^{\mathbf{E}} = \begin{bmatrix} \mathbf{A} & \mathbf{b} & \mathbf{0} \\ \mathbf{0}^{\mathrm{T}} & 1 & 0 \\ \mathbf{c}^{\mathrm{T}} & 0 & 0 \end{bmatrix}, \mathbf{B}^{\mathbf{E}} = \begin{bmatrix} \mathbf{b} & 1 \\ 1 & 1 \\ 0 & 1 \end{bmatrix}, \mathbf{C}^{\mathbf{E}} = \begin{bmatrix} \mathbf{c}^{\mathrm{T}} & 0 & 0 \\ \mathbf{c}^{\mathrm{T}} & 0 & -1 \end{bmatrix}. \tag{6.32}
$$

The second equations in (6.29) and (6.31) can be transformed into the two equivalent expressions (6.33):

$$
\mathbf{e_k} = -\mathbf{C}^{\mathbf{E}}\mathbf{x_k}, \mathbf{x_k} = \mathbf{C}^{\mathbf{b}}\mathbf{e_k} , \tag{6.33}
$$

where the matrix $\mathbf{C}^{\mathbf{b}}$, dim $\mathbf{C}^{\mathbf{b}} = (\mathbf{n}+2, 2)$, can be computed relatively easy as function of $\mathbf{C}^{\mathbf{E}}$.

The core of the proposed stability analysis method can be stated in terms of the following theorem giving sufficient stability conditions.

Theorem 1. The nonlinear system, with the structure presented in Fig. 6.7 (b) and the mathematical model of its linear part (6.29), is globally asymptotically stable if:

– the three matrices $\mathbf{P}$ (positive definite, dim $\mathbf{P} = (\mathbf{n}+2, \mathbf{n}+2)$), $\mathbf{L}$ (regular, dim $\mathbf{L} = (\mathbf{n}+2, \mathbf{n}+2)$) and $\mathbf{V}$ (any, dim $\mathbf{V} = (\mathbf{n}+2, 2)$) fulfill the following requirements:

$$
(\mathbf{A}^{\mathbf{E}})^{\mathrm{T}} \cdot \mathbf{P} \cdot \mathbf{A}^{\mathbf{E}} = -\mathbf{L} \cdot \mathbf{L}^{\mathrm{T}} ,
$$
$$
\mathbf{C}^{\mathbf{E}} - (\mathbf{B}^{\mathbf{E}})^{\mathrm{T}} \cdot \mathbf{P} \cdot \mathbf{A}^{\mathbf{E}} = \mathbf{V}^{\mathrm{T}} \cdot \mathbf{L}^{\mathrm{T}} , \tag{6.34}
$$
$$
-(\mathbf{B}^{\mathbf{E}})^{\mathrm{T}} \cdot \mathbf{P} \cdot \mathbf{B}^{\mathbf{E}} = \mathbf{V}^{\mathrm{T}} \cdot \mathbf{V} ;
$$

– by introducing the matrices $\mathbf{M}$ (dim $\mathbf{M} = (2,2)$), $\mathbf{N}$ (dim $\mathbf{N} = (2,2)$) and $\mathbf{R}$ (dim $\mathbf{R} = (2,2)$) defined as follows:

$$
\mathbf{M} = (\mathbf{C}^{\mathbf{b}})^{\mathrm{T}}(\mathbf{L} \cdot \mathbf{L}^{\mathrm{T}} - \mathbf{P})\mathbf{C}^{\mathbf{b}} ,
$$
$$
\mathbf{N} = (\mathbf{C}^{\mathbf{b}})^{\mathrm{T}}[\mathbf{L} \cdot \mathbf{V} - (\mathbf{A}^{\mathbf{E}})^{\mathrm{T}} \cdot \mathbf{P} \cdot \mathbf{B}^{\mathbf{E}} - 2(\mathbf{C}^{\mathbf{E}})^{\mathrm{T}}] , \tag{6.35}
$$
$$
\mathbf{R} = \mathbf{V}^{\mathrm{T}} \cdot \mathbf{V} ;
$$

the Popov-type inequality (6.36) holds for any value of the control error e_k:

$$
f(\mathbf{e_k}) \cdot \mathbf{n}^{\mathrm{T}} \cdot \mathbf{e_k} + (\mathbf{e_k})^{\mathrm{T}} \cdot \mathbf{M} \cdot \mathbf{e_k} \geq 0 , \tag{6.36}
$$

where $\mathbf{n}$ represents the first column in $\mathbf{N}$.

The proof of Theorem 1, based on Kalman-Szego lemma [19] and on the processing of the Popov sum, is presented in [20] for the PI-FCs with prediction.

By taking into account these aspects, the SAM dedicated to the FCSs comprising PI-fuzzy controllers, consists of the following steps:

step 1: express the MM of the CP, choose the sampling period T_s and compute
the discrete-time state-space mathematical model of the CP with the zero-
order hold, (6.26);

step 2: obtain the discrete-time state-space mathematical model of the ECP,
(6.31);

step 3: compute the matrix $\mathbf{C^b}$ in terms of (6.33);

step 4: solve the system of (6.34), with the solutions $\mathbf{P}$, $\mathbf{L}$ and $\mathbf{V}$, and compute
the matrices $\mathbf{M}$, $\mathbf{N}$ and $\mathbf{R}$ in (6.35);

step 5: choose the value of the free parameter $B_e > 0$ of the PI-FC, and tune
the other parameters of the PI-FC in terms of (6.25);

step 6: check the stability condition (6.36) for any values of the PI-FC inputs corresponding to the operating regimes considered to be significant
regarding the FCS behavior.

To test the presented SAM it has been considered again the case study
corresponding to the EHS control. In this case the controlled plant is represented by the stabilized electro-hydraulic servo-system (SEHS), and the FCS
structure is presented in Fig. 6.9.

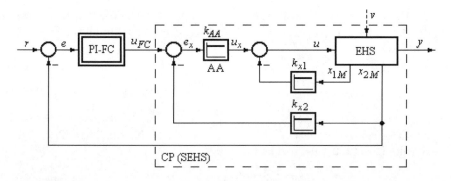

Fig. 6.9. Structure of FCS dedicated to SEHS control

The SEHS represents itself a state feedback control system, with AA –
the adder amplifier, and k_{x1}, k_{x2}, k_{AA} – the parameters of the state feedback
controller. By omitting the nonlinearities of the EHS (Fig. 6.2), imposing the
double pole of the SEHS in -1, the pole placement method leads to the SEHS
transfer function, $H_{CP}(s)$:

$$H_{CP}(s) = \frac{\frac{1}{k_{x2}}}{1 + \frac{k_{M1}T_{i2}k_{x1}}{k_{AA}k_{M2}k_{x2}}s + \frac{T_{i1}T_{i2}}{g_0 k_{AA}k_{M2}}s^2} = \frac{1}{(1+s)^2}, \quad (6.37)$$

obtained for $k_{x1} = 0.2997$, $k_{x2} = 1$, $k_{AA} = 0.0708$.

The steps of the SAM were proceeded but only the values of the matrices
$\mathbf{M}$ and $\mathbf{n^T}$ will be presented because these two matrices appear in the stability
condition (6.36), tested by digital simulation.

$$\mathbf{M} = \begin{bmatrix} 2.0071 & 0 \\ 0 & 0 \end{bmatrix}, \mathbf{n}^{\mathrm{T}} = [\mathbf{0.9855, 0}] , \tag{6.38}$$

and the parameters of the PI-fuzzy controller that ensure the stability of the FCS are tuned as: $B_e = 0.3$, $B_{\triangle e} = 0.0076$, $B_{\triangle u} = 0.0203$. To verify the stability of all FCS the dynamic behavior of the free control system was simulated, when the system started from two different, arbitrarily chosen, initial states, obtained by feeding a $d_3 = -0.5$ and a $d_3 = -1.5$ disturbance input according to Fig. 6.5 (a).

The FCS behavior is presented in Fig. 6.10, and it illustrates that the FCS analyzed by using the presented SAM is stable.

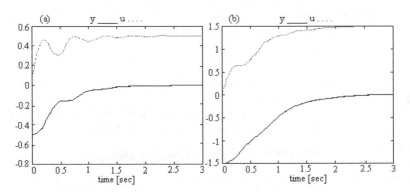

Fig. 6.10. FCS behavior for $d_3 = -0.5$ (**a**) and $d_3 = -1.5$ (**b**)

6.4 Stability Analysis Method Based on the Circle Criterion

The SAM based on the circle criterion refers to the FCS structure presented in Fig. 6.7 (a), with the controlled plant (CP) instead of the block ECP, $\dim \mathbf{r} = \dim \mathbf{y} = \dim \mathbf{e} = (\mathbf{q}, \mathbf{1})$ and $\dim \mathbf{u} = (\mathbf{p}, \mathbf{1})$. It is accepted that the continuous-time CP is a linearized time-invariant n-th order system characterized by the state MM:

$$\dot{\mathbf{x}} = \mathbf{A} \cdot \mathbf{x} + \mathbf{B} \cdot \mathbf{u} , \tag{6.39}$$

$$\mathbf{y} = \mathbf{C} \cdot \mathbf{x} + \mathbf{D} \cdot \mathbf{u} ,$$

where $\dim \mathbf{A} = (\mathbf{n}, \mathbf{n})$, $\dim \mathbf{B} = (\mathbf{n}, \mathbf{p})$, $\dim \mathbf{C} = (\mathbf{q}, \mathbf{n})$ and $\dim \mathbf{D} = (\mathbf{q}, \mathbf{p})$.

The block FC in Fig. 6.7 (a) ensures the nonlinear input-output static map $\mathbf{F}$, $\mathbf{F} : \Re^{\mathbf{q}} \to \Re^{\mathbf{p}}$.

It must be outlined that in the accepted FCS structure it is performed a clear separation between the linear (linearized) and the nonlinear parts of the FCS, which involves a strictly speaking fuzzy controller, without dynamics. If the block FC contains (linear) dynamics, this must be transferred from the block FC to the block CP (see the approach in the previous Section).

The general results of the circle criterion can be applied in the case of FCSs [3,4,8], and grouped under the form of the following lemma that offers sufficient conditions for the asymptotic stability of FCSs.

Lemma 1. To ensure the asymptotic stability of the FCS with the structure in Fig. 6.7 (a) and the state MM (6.39) of the CP it is necessary that:

– the linear part of the FCS (this means the CP) should be stable;
– the conditions (6.40) should be fulfilled:

$$\|\mathbf{H}(\mathbf{w})\| < \frac{1}{K}, \forall \mathbf{w} \in \Re, \|\mathbf{F}(\mathbf{e})\| \leq \mathbf{K}\|\mathbf{e}\|, \forall \mathbf{e} \in \Re^{\mathbf{q}}, \qquad (6.40)$$

where $\mathbf{H}(\mathbf{w})$ – the transfer matrix of the CP and $K = const > 0$.

In the case of SISO controlled plants ($p = q = 1$) the second condition in (6.40) is transformed into the symmetric sector (6.41):

$$-K \cdot \mathrm{sgn}(e) \leq \phi(e) \leq K \cdot \mathrm{sgn}(e), \quad if \quad e \neq 0, \phi(0) = 0, \qquad (6.41)$$

where ϕ is the nonlinear input-output static map of the FC (without dynamics) which provides the control signal $u = \phi(e)$.

Since in the case of linear time-invariant systems the Euclidean norm is equivalent to the H^∞ one [21], the following relation holds:

$$\|\mathbf{H}(\mathbf{w})\| = \|\mathbf{H}(\mathbf{j}\omega)\|_\infty = \mathbf{sup}_{\omega \geq 0}|\mathbf{H}(\mathbf{j}\omega)|. \qquad (6.42)$$

Therefore, by using the first condition in (6.40) and the (6.42), the result can be expressed as:

$$|H(j\omega)| < \frac{1}{K}, \forall \omega \geq 0, \qquad (6.43)$$

which justifies the circle interpretation from a geometrical point of view.

On the basis of a conform transform applied to (6.43) and of the generalization of (6.41) another formulation of the circle criterion can be expressed in terms of Lemma 2.

Lemma 2. To ensure the asymptotic stability of the FCS with the structure in Fig. 6.7 (a) and the state MM (6.39) of the CP it is necessary that:

• the linear part of the FCS (this means the CP) should be stable;
• the FC without dynamics fulfills the conditions (6.44):

$$K_1 e \leq \phi(e) \leq K_2 e \quad if \quad e > 0, K_2 e \leq \phi(e) \leq K_1 e \quad if \quad e < 0, K_2 > K_1; \qquad (6.44)$$

• the linear part of the FCS fulfills the conditions (6.45) and (7.46):

$$|H(j\omega) - C| > R_0, \forall \omega \geq 0, \quad if \quad K_1 K_2 > 0,$$
$$|H(j\omega) - C| < R_0, \forall \omega \geq 0, \quad if \quad K_1 < 0 < K_2; \qquad (6.45)$$
$$Re\, H(j\omega) > -\frac{1}{K_2}, \forall \omega \geq 0, \quad if \quad K_1 = 0, K_2 > 0;$$
$$Re\, H(j\omega) < -\frac{1}{K_1}, \forall \omega \geq 0, \quad if \quad K_1 < 0, K_2 = 0. \qquad (6.46)$$

The circle centered on the real axis of the complex plane with the equation $|H(j\omega) - C| = R_0$ which appears in (6.45) is characterized by the abscissa of its center C and the radius R_0:

$$C = -0.5 \left(\frac{1}{K_1} + \frac{1}{K_2} \right), R_0 = 0.5 \left| \frac{1}{K_1} - \frac{1}{K_2} \right|. \tag{6.47}$$

The SAM in terms of the circle criterion, dedicated to FCSs, consists of the following steps:

step 1: use (6.44) to determine the constants K_1 and K_2, $K_1 < K_2$, which ensure the limitation of the input-output static map of the FC;

step 2: represent the Nyquist plot associated to $H(j\omega)$ for $\omega \geq 0$;

step 3: check one of the following four conditions:

1. the Nyquist plot is placed entirely in the external part of the circle $|s - C| = R_0$, with s – the complex variable, $s = \sigma + j\omega$, if $K_1 K_2 > 0$ (the first equation in (6.45)), or

2. the Nyquist plot is placed entirely in the internal part of the circle $|s - C| = R_0$, if $K_1 < 0 < K_2$ (the second equation in (6.45)), or

3. the Nyquist plot is placed entirely in the right-hand side of the line $s = -\frac{1}{K_2}$, if $K_1 = 0$, $K_2 > 0$ (the first equation in (6.46)), or

4. the Nyquist plot is placed entirely in the right-hand side of the line $s = -\frac{1}{K_1}$, if $K_1 < 0$, $K_2 = 0$ (the second equation in (6.46));

the fulfillment of one of the conditions a) $\cdots$ b) guarantees the asymptotic stability of the FCS.

The graphical interpretation of the sufficient stability conditions (6.44) $\cdots$ (6.46) is presented in Fig. 6.11.

The application of the presented SAM is done for the FCS with the structure presented in Fig. 6.11 (which is a modified structure with respect to that in Fig. 6.9). The FC is developed to replace the block AA (in Fig. 6.9) for performance enhancement, but in this case the reference input is r_x taking the place of u_{FC}.

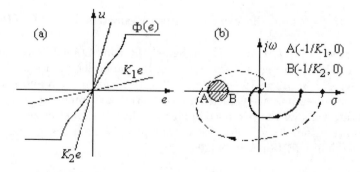

Fig. 6.11. Interpretation of the stability conditions (44) (a) and (45), (46) (b)

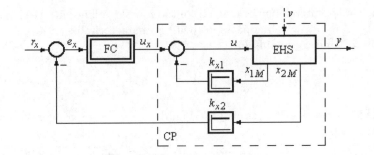

Fig. 6.12. Structure of FCS

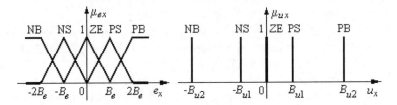

Fig. 6.13. Input and output membership functions of FC in Fig. 6.12

The FC, which is without dynamics, in Fig. 6.12 is characterized by the input and output membership functions illustrated in Fig. 6.13, it employs Mamdani's MAX-MIN compositional rule of inference assisted by the rule base presented in Table 6.3, and the center of gravity method for defuzzification.

Table 6.3. Decision table of FC in Fig. 6.12

step 4:

e_x	NB	NS	ZE	PS	PB
u_x	NB	NS	ZE	PS	PB

Fig. 6.13 points out the strictly positive parameters of the FC, $\{B_e, B_{u1}, B_{u2}\}$.

To develop the FC it is firstly designed the conventional state feedback controller (SEHS) in Fig. 6.9. By imposing the closed-loop poles in the form of $p_1^* = -p + j \cdot q$, $p_2^* = -p - j \cdot q$ $(p > 0, q \geq 0)$, for $p = 3.5$ and $q = 6.05$, applying the pole placement method (see (6.37)) and imposing that for the steady-state value of the reference input $r_x = 10V$ it should be obtained the steady-state value of the controlled output $y = 200\,\text{mm}$, the parameters of the conventional SEHS result as [20]:

$$k_{x1} = \frac{2T_{i1}p}{g_0 k_{M1}} = 1.05, k_{x2} = \frac{r_x}{k_{M2}y} = 1.56, k_{AA} = \frac{(p^2 + q^2)T_{i1}T_{i2}}{g_0 k_{AA} k_{M2}} . \quad (6.48)$$

From the point of view of the FC development, this SEHS will operate for big absolute values of the control error e_x. For small absolute values of e_x it is developed another conventional SEHS, which ensures the oscillations suppression ($q = 0$) resulting in the parameters k_{x1} and k_{x2} presented in (6.48), but with $k_{AA} = 0.56$.

The parameters of the FC are obtained by applying the modal equivalence principle [12] in the accepted conditions: $B_e = 0.5$, $B_{u1} = 0.28$ and $B_{u2} = 2.2$.

By applying the SAM for the FCS in Fig. 6.12, the limits of the input-output static map of the FC are $K_1 = 0.05$ and $K_2 = 0.8$. The Nyquist plot associated to $H(j\omega)$ for $\omega \geq 0$ has the shape presented in Fig. 6.11 (b) with continuous line, and this proves that the considered FCS is asymptotically stable.

6.5 Harmonic Balance Approach to Stability Analysis

This SAM, dedicated to the stability analysis of nonlinear dynamic systems [22], can be applied to the stability analysis of the FCSs with the structure presented in Fig. 6.7 (a), with the controlled plant (CP) instead of the block ECP. But, the controlled plant should be a single input system ($p = 1$), so the control signal vector **u** will be replaced by the control signal denoted by the scalar u.

For the stability analysis of the EP placed in the origin of the state-space it is necessary that $\mathbf{r_k} = \mathbf{0}$ and the eventual disturbance inputs should be also zero. Furthermore, the following conditions must be fulfilled:

(a) The static nonlinearity of the FC should be symmetrical.
(b) The CP should have Bode plots of low pass filter type.

If it is interrupted the connection between the FC and the CP it is obtained the open-loop system, illustrated in Fig. 6.14–(a), corresponding to the FCS in Fig. 6.14–(b); the open-loop system admits as input u and as output $\hat{u}$. So, the third necessary condition to be fulfilled by the FCS for the application of the harmonic balance method is:

(c) The open-loop system should be stable.

In the conditions (a) $\cdots$ (c) it can be considered [10,23] that for a nonlinear system, by applying a harmonic input (6.49):

$$U(A, j\omega) = A \cdot e^{j\omega t} , \tag{6.49}$$

with the magnitude $A > 0$ and the frequency $\omega \geq 0$, it will be obtained at the system output a periodic non-sinusoidal signal. By neglecting the higher-order harmonics in the Fourier series of this signal, the fundamental harmonics of the output signal is obtained in terms of (6.50):

$$\hat{U}(A, j\omega) = |\hat{U}(A, j\omega)| \cdot e^{j(\omega t + \varphi)} = |\hat{U}| \cdot e^{j(\omega t + \varphi)} . \tag{6.50}$$

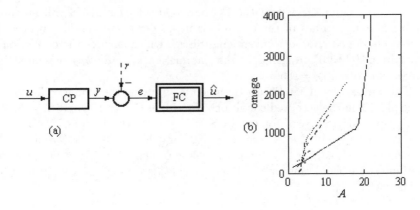

Fig. 6.14. Open-loop system structure (**a**); interpretation in the case study (**b**)

The open-loop system behavior can be characterized by its associated describing function, $N(A, j\omega)$, defined as follows:

$$N(A, j\omega) = \frac{\hat{U}(A, j\omega)}{U(A, j\omega)} .$$ (6.51)

By using (6.49) and (6.50), the describing function can be expressed as:

$$N(A, j\omega) = \frac{|\hat{U}|}{A} \cdot e^{j\varphi} ,$$ (6.52)

and the harmonic balance condition is given by (6.53):

$$N(A, j\omega) = 1 .$$ (6.53)

The pairs of values (A, ω) for which the condition (6.53) is fulfilled point out the fact that in the closed-loop system (the FCS) there exist permanent oscillations of magnitude A and frequency ω. If A increases, then the modulus of the describing function, $|N(A, j\omega)|$, will decrease and this decrease is illustrated in the plane $\langle A, \omega \rangle$ by points (A, ω) placed in the vicinity of the points that fulfill the harmonic balance condition (6.53), with the result in stable permanent oscillations and vice versa.

The analytical check of the condition (6.53) in fuzzy control applications is relatively difficult. This is the reason why it is simpler to calculate point by point the values of the describing function, $N(A, j\omega)$, on the basis of the digital simulation of the open-loop system behavior for a set of M pairs of values $(A_i, \omega_i), i = 1 \cdots M$.

These aspects result in a specific formulation of the SAM based on the harmonic balance method. The SAM consists of the following steps:
• Step 1: Feed the function (6.54) to the open-loop system:

$$u(t) = Im\, U(A_i, j\omega_i) = A_i \sin(\omega_i t), i = 1 \cdots M \ . \tag{6.54}$$

• Step 2: After the system enters the oscillatory permanent regime record the values of the output function, $u_i(t), i = 1 \cdots M$. These values are approximated analytically by the following function:

$$u_i(t) \simeq A_i' \sin(\omega_i t) + B_i' \cos(\omega_i t), i = 1 \cdots M \ . \tag{6.55}$$

The values of the parameters A_i' and B_i' are determined by applying the least squares (LS) method [24] on the basis of a set of P measurements for different time moments t_k, resulting in the pairs of values $(\omega_i t_k, u_{ik}), k = 1 \cdots P$, where $u_{ik} = u_i(t_k)$. The LS method requires the minimization of the cost function J for a given index i:

$$J = f_1(A_i', B_i') = \sum_{k=1}^{P} [u_{ik} - A_i' \sin(\omega_i t) - B_i' \cos(\omega_i t)]^2 \ . \tag{6.56}$$

By introducing the matrices $\mathbf{y}^*$, $\mathbf{S}$ and $\mathbf{p}$ *:

$$\mathbf{y}^* = \begin{bmatrix} u_{i1} \\ u_{i2} \\ \cdots \\ u_{iP} \end{bmatrix}, \mathbf{S} = \begin{bmatrix} \sin(\omega_i t_1) & \cos(\omega_i t_1) \\ \sin(\omega_i t_2) & \cos(\omega_i t_2) \\ \cdots & \cdots \\ \sin(\omega_i t_P) & \cos(\omega_i t_P) \end{bmatrix}, \mathbf{p}^* = \begin{bmatrix} A_i' \\ B_i' \end{bmatrix}, \tag{6.57}$$

the objective function can be expressed in the form (6.58):

$$J = f_2(\mathbf{p}^*) = (\mathbf{y}^* - \mathbf{S} \cdot \mathbf{p}^*)^{\mathbf{T}}(\mathbf{y}^* - \mathbf{S} \cdot \mathbf{p}^*) \ . \tag{6.58}$$

Taking the derivative with respect to $\mathbf{p}^*$ in (6.58) and imposing the zero gradient:

$$\frac{dJ}{d\mathbf{p}^*} = -2\mathbf{S}^{\mathbf{T}} \cdot (\mathbf{y}^* - \mathbf{S} \cdot \mathbf{p}^*) = \mathbf{0} \ , \tag{6.59}$$

results in the value of the stationary point, $\mathbf{p}^*$:

$$\mathbf{p}^* = (\mathbf{S}^{\mathbf{T}} \cdot \mathbf{S})^{-1} \cdot \mathbf{S}^{\mathbf{T}} \cdot \mathbf{y}^* \ . \tag{6.60}$$

This stationary point is an absolute minimum one because the Hessian matrix calculated here is positive definite:

$$\frac{d^2 J}{d\mathbf{p}^{*2}} = 2\mathbf{S}^{\mathbf{T}} \cdot \mathbf{S} > \mathbf{0} \ . \tag{6.61}$$

So, it can be concluded that the values of the parameters A_i' and B_i' are obtained by applying (6.60) M times, $i = 1 \cdots M$.

• Step 3: Obtain the modulus and the argument of the describing function as functions of the parameters A_i' and B_i' in terms of (6.62):

$$|N(A_i, j\omega_i)| = \frac{\sqrt{A_i'^2(A_i) + B_i'^2(A_i)}}{A_i} , \tag{6.62}$$

$$\varphi = \arg\{N(A_i, j\omega_i)\} = \arg\{A_i'(A_i) + j \cdot B_i'(A_i)\}, i = 1 \cdots M .$$

• Step 4: To find the points (A_i, ω_i) that fulfill the harmonic balance condition (6.53), interpret the describing function $N(A, j\omega)$ as a vector in the plane $<A, \omega>$. For this aim there are calculated by interpolation and represented in the plane $<A, \omega>$ the following lines with the parameter ε:

$$\varphi = N(A, j\omega) = 0, |N(A, j\omega)| - 1 = \varepsilon . \tag{6.63}$$

The second family of lines in (6.63) consists of three level lines obtained for $\varepsilon = 0$, $\varepsilon > 0$ and $\varepsilon < 0$. The non-zero values of ε must be chosen to take relatively small absolute values.

If for $\varepsilon = 0$ the lines (6.63) intersect in a point (A^*, ω^*), then the harmonic balance condition will be fulfilled, which will indicate the presence of a permanent oscillation in the system.

• Step 5: If in the vicinity of the point (A^*, ω^*) obtained in the previous step it is observed that the intersection of the lines (6.63) for $\varepsilon > 0$ occurs for $A < A^*$, and the intersection for $\varepsilon < 0$ occurs for $A > A^*$, then the permanent oscillation will be stable, and the FCS will be stable.

It should be pointed out finally that the major difficulty in applying the method concerns the proper choice of the domain of frequencies ω in which there are performed the digital simulations as part of the step 1.

To apply this SAM it is accepted the case study presented in Sect. 6.4. Proceeding the steps $(1) \cdots (5)$ will give the situation illustrated in Fig. 6.14 (b), where there are represented in the plane $<A, \omega>$ the following lines: the first line in (6.63) with continuous line $(- - -)$, the second line in (6.63) for $\varepsilon = 0$ with dash dotted line $(-. - .-)$, the second line in (6.63) for $\varepsilon = 0.05$ with dotted line $(\cdots)$, and the second line in (6.63) for $\varepsilon = -0.05$ with dashed line $(- - -)$. All lines were obtained by interpolation starting with 40 points obtained by digital simulation.

6.6 Sensitivity Analysis of a Class of Fuzzy Control Systems

Let the considered control system structure be a conventional one, presented in Fig. 6.15 (a), where: C – the controller, CP – the controlled plant, RF – the reference (feedforward) filter, r – the reference input, r – the filtered reference input, e – the control error, u – the control signal, y – the controlled output, d_1, d_2, d_3, d_4 – disturbance inputs.

Depending on the place of feeding the disturbance inputs to the CP and on the CP structure, the accepted types of disturbance inputs $\{d_1, d_2, d_3, d_4\}$ are defined in terms of Fig. 6.15 (b).

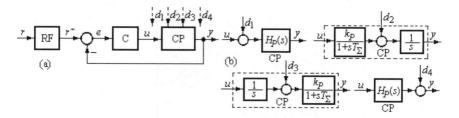

Fig. 6.15. Control system structure (**a**); disturbance inputs types (**b**)

The class of plants with the simplified structure illustrated in Fig. 6.15 (a), is considered to be linearized around a steady-state operating point and characterized by the transfer function $H_P(s)$:

$$H_P(s) = \frac{k_P}{s(1 + T_\Sigma s)} \,, \tag{6.64}$$

with k_P – gain and T_Σ – small time constant or sum of all parasitic time constants, belongs to a class of integral-type systems with variable parameters applicable to servo-systems.

For these plants, it is recommended the use of linear PI controllers having the transfer function $H_C(s)$ (see also (6.23)):

$$H_C(s) = k_C \left(1 + \frac{1}{T_i s}\right) = k_c \frac{1 + T_i s}{s}, k_C = k_c T_i \,. \tag{6.65}$$

Based on the Extended Symmetrical Optimum (ESO) method [25], the parameters of the controller k_c (or k_C) – controller gain and T_i – integral time constant are tuned on the basis of (6.66) guaranteeing the desired control system performance by means of a design parameter β:

$$k_c = \frac{1}{\beta^{3/2} k_P T_\Sigma^2}, T_i = \beta T_\Sigma, k_C = \frac{1}{\beta^{1/2} k_P T_\Sigma} \,. \tag{6.66}$$

The tuning relations (6.66) ensure good Control System Performance Indices (CSPIs) σ_1 – overshoot, t_s – settling time, t_1 – first settling time, and a maximum phase margin (φ_r). A minimum guaranteed phase margin can be achieved for plants with time-varying k_P.

By using (6.66), the closed-loop transfer function with respect to the filtered reference input, $H_r(s)$, will be expressed as:

$$H_r(s) = \frac{\beta T_\Sigma s + 1}{\beta^{3/2} T_\Sigma^3 s^3 + \beta^{3/2} T_\Sigma^2 s^2 + \beta T_\Sigma s + 1} \,. \tag{6.67}$$

By the choice of the parameter β in the recommended domain $1 < \beta < 20$, the CSPIs $\{\sigma_1, t_s^E = t_s/T_\Sigma, t_1^E = t_1/T_\Sigma, \varphi_r\}$ can be accordingly modified

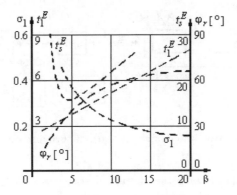

Fig. 6.16. Control system performance indices versus β

and a compromise to these performance indices can be reached by using the diagrams illustrated in Fig. 6.16.

It can be observed that all coefficients in $H_r(s)$ depend on the parameter β and that the PI controller development is reduced to the choice of this single parameter, β. The reference filters can further improve the CSPIs by the cancellation of the zero or of the pair of complex conjugated poles in (6.67).

As it has been shown in [26], the variation of CP parameters ($\{k_P, T_\Sigma\}$ for the considered CPs) due to the change of the steady-state operating points or to other conditions leads to additional motion (of the control systems). This motion is usually undesirable under uncontrollable parametric variations. Therefore, to alleviate the effects of parametric disturbances it is necessary to perform the sensitivity analysis with respect to the parametric variations of the CP.

It is generally accepted that FCs ensure control system performance enhancement with respect to the modifications of the reference input, of the load disturbance inputs or to parametric variations. This justifies the research efforts focused on the systematic analysis of FCSs behavior with respect to parametric variations and the need to perform the sensitivity analysis with this respect that enables to derive sensitivity models [26] for the FCs and for the overall FCSs.

The sensitivity models enable the sensitivity analysis of the FCSs accepted, as mentioned in Sect. 6.1, to be approximately equivalent with the linear control systems. This justifies the approach to be presented in the sequel, that the sensitivity models of the FCSs are approximately equivalent to the sensitivity models of the linear ones. Therefore, it is necessary to obtain firstly the sensitivity models of the linear control system (Fig. 6.15 (a)).

To derive the sensitivity models of the linear control system with respect to the variations of CP parameters k_P and T_Σ it will be considered the CP structure in Fig. 6.15 corresponding to a d_3 type disturbance input.

By considering the state variables x_1 (the controlled output) and x_2 (the output of the integral element), the state MM of the CP results as follows:

$$\dot{x}_1(t) = -(1/T_\Sigma)x_1(t) + (k_P/T_\Sigma)x_2(t) + (k_P/T_\Sigma)d_3(t) ,$$
$$\dot{x}_2(t) = u(t) ,$$
$$y(t) = x_1(t) . \tag{6.68}$$

The state MM of the linear PI controller can be expressed in its parallel form:

$$\dot{x}_3(t) = (1/T_i)e(t) ,$$
$$u(t) = k_C(x_3(t) + e(t)) , \tag{6.69}$$

where x_3 is the output of the integral component in the controller.

To perform the sensitivity analysis with respect to the parametric variations of the CP, the linear PI controller is tuned in terms of (6.66) by considering the nominal values of controlled plant parameters, $\{k_{P0}, T_{\Sigma 0}\}$. Therefore, the state MM of the PI controller (6.69) will be transformed into (6.70):

$$\dot{x}_3(t) = [1/(\beta T_{\Sigma 0})]e(t) ,$$
$$u(t) = [1/(\beta^{1/2}k_{P0}T_{\Sigma 0})](x_3(t) + e(t)) , \tag{6.70}$$

The state MM of the closed-loop system can be obtained by the merge of the models in (6.68) and (6.70) resulting in:

$$\dot{x}_1(t) = -(1/T_\Sigma)x_1(t) + (k_P/T_\Sigma)x_2(t) + (k_P/T_\Sigma)d_3(t) ,$$
$$\dot{x}_2(t) = -\left[1/(\beta^{1/2}k_{P0}T_{\Sigma 0})\right]x_1(t) + \left[1/(\beta^{1/2}k_{P0}T_{\Sigma 0})\right]x_3(T)$$
$$+ \left[1/(\beta^{1/2}k_{P0}T_{\Sigma 0})\right]r(t) ,$$
$$\dot{x}_3(t) = -\left[1/(\beta T_{\Sigma 0})\right]x_1(t) + [1/(\beta T_{\Sigma 0})]r(t) ,$$
$$y(t) = x_1(t) . \tag{6.71}$$

For the system (6.71) there can be derived the state sensitivity functions λ_1, λ_2, λ_3 and the output sensitivity function, σ [27]:

$$\lambda_j(t) = [\partial x_j(t)/\partial \alpha]_{\alpha 0}, \sigma(t) = [\partial y(t)/\partial \alpha]_{\alpha 0}, j = 1 \cdots 3 , \tag{6.72}$$

where the subscript 0 stands for the nominal values of the controlled plant parameters, $\alpha \in \{k_P, T_\Sigma\}$.

With this respect, the sensitivity analysis can be applied by considering two parametric variations, of k_P and T_Σ, and the dynamic regimes characterized by: the step modification of the reference input r for $d_3(t) = 0$, or the step modification of the disturbance input d_3 for $r(t) = 0$. This leads to four sensitivity models obtained by computing the partial derivatives with respect to k_P and T_Σ, and presented as follows:

– the sensitivity model with respect to the variation of k_P, the step modification of r, and $d_3(t) = 0$:

$$\dot{\lambda}_1(t) = \lambda_2(t) \,,$$
$$\dot{\lambda}_2(t) = -\left[1/\left(\beta^{1/2}T_{\Sigma 0}^2\right)\right]\lambda_1(t) - (1/T_{\Sigma 0})\lambda_2(t) + \left[1/\left(\beta^{1/2}T_{\Sigma 0}^2\right)\right]\lambda_3(t)$$
$$- \left[1/\left(\beta^{1/2}k_{P0}T_{\Sigma 0}^2\right)\right]x_{10}(t) + \left[1/\left(\beta^{1/2}k_{P0}T_{\Sigma 0}^2\right)\right]x_{30}(t)$$
$$+ \left[1/\left(\beta^{1/2}k_{P0}T_{\Sigma 0}^3\right)\right]r_0(t) \,,$$
$$\dot{\lambda}_3(t) = -\left[1/\left(\beta T_{\Sigma 0}\right)\right]\lambda_1(t) \,,$$
$$\sigma(t) = \lambda_1(t); \tag{6.73}$$

– the sensitivity model with respect to the variation of T_Σ, the step modification of r, and $d_3(t) = 0$:

$$\dot{\lambda}_1(t) = \lambda_2(t) \,,$$
$$\dot{\lambda}_2(t) = -\left[1/\left(\beta^{1/2}T_{\Sigma 0}^2\right)\right]\lambda_1(t) - (1/T_{\Sigma 0})\lambda_2(t) + \left[1/\left(\beta^{1/2}T_{\Sigma 0}^2\right)\right]\lambda_3(t)$$
$$+ \left[1/\left(\beta^{1/2}T_{\Sigma 0}^3\right)\right]x_{10}(t) + (1/T_{\Sigma 0}^2)x_{20}(t)$$
$$- \left[1/\left(\beta^{1/2}T_{\Sigma 0}^3\right)\right]x_{30}(t) - \left[1/\left(\beta^{1/2}T_{\Sigma 0}^3\right)\right]r_0(t) \,,$$
$$\dot{\lambda}_3(t) = -\left[1/\left(\beta T_{\Sigma 0}\right)\right]\lambda_1(t) \,,$$
$$\sigma(t) = \lambda_1(t) \,; \tag{6.74}$$

– the sensitivity model with respect to the variation of k_P, the step modification of d_3, and $r(t) = 0$:

$$\dot{\lambda}_1(t) = -(1/T_{\Sigma 0})\lambda_1(t) + (k_{P0}/T_{\Sigma 0})\lambda_2(t) + (1/T_{\Sigma 0})x_{20}(t)$$
$$+ (1/T_{\Sigma 0})d_{30}(t) \,,$$
$$\dot{\lambda}_2(t) = -\left[1/\left(\beta^{1/2}k_{P0}T_{\Sigma 0}\right)\right]\lambda_1(t) + \left[1/\left(\beta^{1/2}k_{P0}T_{\Sigma 0}\right)\right]\lambda_3(t) \,,$$
$$\dot{\lambda}_3(t) = -\left[1/\left(\beta T_{\Sigma 0}\right)\right]\lambda_1(t) \,,$$
$$\sigma(t) = \lambda_1(t) \,; \tag{6.75}$$

– the sensitivity model with respect to the variation of T_Σ, the step modification of d_3, and $r(t) = 0$:

$$\dot{\lambda}_1(t) = -(1/T_{\Sigma 0})\lambda_1(t) + (k_{P0}/T_{\Sigma 0})\lambda_2(t) + (1/T_{\Sigma 0}^2)x_{10}(t)$$
$$- (k_{P0}/T_{\Sigma 0}^2)x_{20}(t) - (k_{P0}/T_{\Sigma 0}^2)d_{30}(t) \,,$$
$$\dot{\lambda}_2(t) = -\left[1/\left(\beta^{1/2}k_{P0}T_{\Sigma 0}\right)\right]\lambda_1(t) + \left[1/\left(\beta^{1/2}k_{P0}T_{\Sigma 0}\right)\right]\lambda_3(t) \,,$$
$$\dot{\lambda}_3(t) = -\left[1/\left(\beta T_{\Sigma 0}\right)\right]\lambda_1(t) \,,$$
$$\sigma(t) = \lambda_1(t) \,. \tag{6.76}$$

It must be highlighted that in the sensitivity models (6.73) $\cdots$ (6.76) there have been used the notations: $\{x_{10}, x_{20}, x_{30}\}$ – the nominal values of the state variables, r_0 – the nominal value of the reference input, and d_{30} – the nominal value of the disturbance input. These variables determine the nominal trajectory of the control system or its fundamental motion.

The derived sensitivity models (6.73) $\cdots$ (6.76) can be considered as valid for both the linear control system with PI controller (the structure in Fig. 6.15 (a)) and the FCS with PI-fuzzy controller (replacing the block C in Fig. 6.15 (a)) presented in Sect. 6.3 and developed by the modal equivalence principle [12]. This is justified due to the approximate equivalence mentioned in Sect. 6.1 between the fuzzy control systems and the linear ones. However, the difference between the sensitivity models of the FCS and of the linear control system is in the generation of the nominal trajectory of the control system by using a PI-FC or a linear PI controller, respectively.

To solve the control problems in mobile robots defined in [28], trajectory tracking, path following and point stabilization, there are widely used several mathematical models, including the kinematical ones [29] and the dynamic ones [30, 31]. Since the model represented by the transfer function in (6.63) appears in more complex or simpler forms of servo-systems in these dynamic models, the robot control problems can be simulated by considering this model as benchmark [32].

In this context, for the validation of the proposed sensitivity models it is considered a case study with the CP characterized in its linearized simplified form by the transfer function (6.64), with the nominal values of CP parameters $k_{P0} = 1$ and $T_{\Sigma 0} = 1sec$. The control system structure is the conventional one presented in Fig. 6.15 (a), with the PI-FC having the structure presented in Sect. 6.3 as controller (C).

For the development of the PI-FC in the considered case study it is applied the ESO method, which starts with choosing $\beta = 6$, and the parameters of the PI controller will obtain the nominal values $k_{c0} = 0.068$, and $T_{i0} = 6sec$. Then, by proceeding with the development of the PI-FC in terms of Sect.s 6.3, the values of the tuning parameters of the PI-FC result as: $B_e = 0.3$, $B_{\triangle e} = 0.03$, $B_{\triangle u} = 0.0021$.

To conclude the sensitivity analysis concerning the developed FCS there are presented in Fig. 6.17 the behaviors of the sensitivity models (6.73) $\cdots$ (6.76), obtained for the simulation scenario that employs a unit step modification of r followed by a unit step modification of d_3 (after 250 sec) in the initial conditions $\lambda_1(0) = 2$, $\lambda_2(0) = 1$, $\lambda_3(0) = 0$:

– in Fig. 6.17 (a): the behavior of the sensitivity model in (6.73), obtained with respect to the variation of k_P, the step modification of r, and $d_3(t) = 0$;
– in Fig. 6.17 (b): the behavior of the sensitivity model in (6.74), obtained with respect to the variation of T_Σ, the step modification of r, and $d_3(t) = 0$;
– in Fig. 6.17 (c): the behavior of the sensitivity model in (6.75), obtained with respect to the variation of k_P, the step modification of d_3, and $r(t) = 0$;
– in Fig. 6.17 (d): the behavior of the sensitivity model in (6.76), obtained with respect to the variation of T_Σ, the step modification of d_3, and $r(t) = 0$.

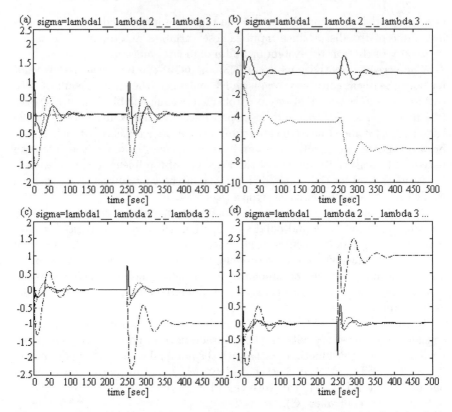

Fig. 6.17. Sensitivity functions versus time for the sensitivity models (73) (in (**a**)), (74) (in (**b**)), (75) (in (**c**)) and (76) (in (**d**))

Figure 6.17 illustrates the sensitivity functions $\sigma(t) = \lambda_1(t)$, (with continuous line, $---$), $\lambda_2(t)$ (with dash dotted line, $-.--.-$) and $\lambda_3(t)$ (with dotted line, $\cdots$).

Although the behaviors of the sensitivity models prove to be good, they can be improved in the presence of the reference filter. Another way to improve the sensitivity of the FCSs with respect to the parameter variations of the controlled plant is to construct optimal FCSs developed by the minimization of several optimization criteria used as quadratic performance indices. These indices employ the output sensitivity function and a development method for FCSs with Takagi-Sugeno PI-fuzzy controllers that the minimization of ten quadratic performance indices was proposed in [33].

6.7 Summary

The chapter presents attractive stability analysis methods and performs the sensitivity analysis, with respect to the parametric variations of the controlled plant, of fuzzy control systems with Mamdani fuzzy controllers dedicated to servo-systems control.

There are offered four stability analysis methods from a relatively wide area of stability analysis methods generally used in the case of fuzzy control systems [2–4]: the state-space approach, the use of Popov's hyperstability theory, the circle criterion and the harmonic balance method. The stability analysis methods are formulated in analysis steps relatively easy to be performed by the practitioners.

In addition, by accepting that in certain conditions the fuzzy control systems are equivalent to the linear ones, there are derived sensitivity models that enable the sensitivity analysis of a class of fuzzy control systems with respect to the parametric variations of the controlled plant.

The case studies presented in the chapter correspond to electro-hydraulic servo-systems and to servo-systems used in control problems concerning mobile robots. These case studies validate the stability and sensitivity analyses.

All presented methods are relatively general and simple, and can be extended with no major supplementary difficulties to other fuzzy controllers including the adaptive fuzzy controllers employing Takagi-Sugeno fuzzy controllers [34].

By presenting the stability and sensitivity analysis methods there are offered useful design recommendations for the fuzzy controllers with dynamics resulting in low-cost fuzzy controllers in comparison with other controllers meant for mobile robots belonging to the general class of controlling nonholonomic or nonsmooth systems [35, 36].

References

1. L.T. Koczy, Fuzzy If ··· Then Rule Models and Their Transformation into One Another, IEEE Trans. Systems, Man, and Cybernetics – Part A, vol. 26, pp. 621–637, 1996
2. M. Sugeno, On Stability of Fuzzy Systems Expressed by Fuzzy Rules with Singleton Consequents, IEEE Trans. Fuzzy Systems, vol. 7, pp. 201–224, 1999
3. D. Driankov, H. Hellendoorn, and M. Reinfrank, An Introduction to Fuzzy Control, Springer-Verlag, Berlin, Heidelberg, New York, 1993
4. K.M. Passino and S. Yurkovich, Fuzzy Control, Addison Wesley Longman, Inc., Menlo Park, CA, 1998
5. J. Aracil, A. Ollero, and A. Garcia-Cerezo, Stability Indices for the Global Analysis of Expert Control Systems, IEEE Trans. Systems, Man, and Cybernetics, vol. 19, pp. 998–1007, 1989
6. A. Garcia-Cerezo and A. Ollero, Stability of Fuzzy Control Systems by Using Nonlinear System Theory, Proceedings of IFAC/IFIP/IMACS Symposium on Artificial Intelligence in Real-Time Control, Delft, pp. 171–176, 1992
7. R.-E. Precup, S. Doboli, and S. Preitl, Stability Analysis and Development of a Class of Fuzzy Control Systems, Engineering Applications of Artificial Intelligence, vol. 13, pp. 237–247, 2000

8. H.-P. Opitz, Fuzzy Control and Stability Criteria, Proceedings of First European Congress on Fuzzy and Intelligent Technologies – EUFIT'93, Aachen, vol. 1, pp. 130–136, 1993

9. R.-E. Precup and S. Preitl, Popov-type Stability Analysis Method for Fuzzy Control Systems, Proceedings of Fifth European Congress on Intelligent Technologies and Soft Computing – EUFIT'97, Aachen, vol. 2, pp. 1306–1310, 1997

10. H. Kiendl, Harmonic Balance for Fuzzy Control Systems, Proceedings of First European Congress on Fuzzy and Intelligent Technologies – EUFIT'93, Aachen, vol. 1, pp. 137–141, 1993

11. K.L. Tang and R.J. Mulholland, Comparing Fuzzy Logic with Classical Controller Designs, IEEE Trans. Systems, Man, and Cybernetics, vol. 17, pp. 1085–1087, 1987

12. S. Galichet and L. Foulloy, Fuzzy Controllers: Synthesis and Equivalences, IEEE Trans. Fuzzy Systems, vol. 3, pp. 140–148, 1995

13. B.S. Moon, Equivalence between Fuzzy Logic Controllers and PI Controllers for Single Input Systems, Fuzzy Sets and Systems, vol. 69, pp. 105–113, 1995

14. K.J. Astrom and T. Hagglund, PID Controllers Theory: Design and Tuning, Instrument Society of America, Research Triangle Park, NC, 1995

15. S. Preitl and D. Onea, Analytical and Experimental Identification of Electrohydraulic Speed Controllers Meant for Hydro-generators (in Romanian), Scientific and Technical Bulletin of I. P. T. V.T. Series Electrical Engineering, vol. 26 (40), pp. 83–93, 1981

16. S. Preitl and R.-E. Precup, Introduction to Fuzzy Control (in Romanian), Editura Tehnica, Bucharest, 1997

17. D.J. Hill and C.N. Chong, Lyapunov Functions of Lure-Postnikov Form for Structure Preserving Models of Power Plants, Automatica, vol. 25, pp. 453–460, 1989

18. K.J. Astrom and B. Wittenmark, Computer-Controlled Systems: Theory and Design, Third ed., Prentice-Hall, Inc., Upper Saddle River, NJ, 1997

19. I.D. Landau, Adaptive Control, Marcel Dekker, Inc., New York, 1979

20. R.-E. Precup S. Preitl, and G. Faur, PI Predictive Fuzzy Controllers for Electrical Drive Speed Control: Methods and Software for Stable Development, Computers in Industry, vol. 52, pp. 253–270, 2003

21. P. Dorato, C.-L. Shen, and W. Yang, Robust Control Systems Design, China Aviation Industry Press, Beijing, 1996

22. A. Isidori, Nonlinear Control Systems: An Introduction, Springer-Verlag, Berlin, Heidelberg, New York, 1989

23. F. Gordillo, F. Salas, R. Ortega, and J. Aracil, Hopf Bifurcation in Indirect Field-oriented Control of Induction Motors, Automatica, vol. 38, pp. 829–835, 2002

24. L. Ljung, System Identification – Theory for the User, Englewood Cliffs, NJ, 1987

25. S. Preitl and R.-E. Precup, An Extension of Tuning Relations after Symmetrical Optimum Method for PI and PID Controllers, Automatica, vol. 35, pp. 1731–1736, 1999

26. F. Rosenwasser and R. Yusupov, Sensitivity of Automatic Control Systems, CRC Press, Boca Raton, FL, 2000

27. P.-M. Frank, Introduction to System Sensitivity Theory, Academic Press, New York, 1978

28. R. Fierro and F.L. Lewis, Control of a Nonholonomic Mobile Robot Using Neural Networks, IEEE Transactions on Neural Networks, vol. 9, pp. 589–600, 1998
29. Y. Kanayama, Y. Kimura, F. Miyazaki, and T. Noguchi, A Stable Tracking Control Method for an Autonomous Mobile Robot, Proceedings of IEEE Conference on Robotics and Automation, Cincinnati, OH, pp. 384–389, 1990
30. K. Watanabe, J. Tang, M. Nakamura, S. Koga, and T. Fukuda, A Fuzzy-Gaussian Neural Network and Its Application to Mobile Robot Control, IEEE Trans. Control Systems Technology, vol. 2, pp. 193–199, 1996
31. Z.-P. Jiang and H. Nijmeijer, Tracking Control of Mobile Robots: A Case Study in Backstepping, Automatica, vol. 33, pp. 1393–1399, 1997
32. R.-E. Precup, S. Preitl, C. Szabo, Z. Gyurko, and P. Szemes, Sliding Mode Navigation Control in Intelligent Space, Proceedings of 2003 International Symposium on Intelligent Signal Processing – WISP 2003, Budapest, pp. 225–230, 2003
33. R.-E. Precup and S. Preitl, Multiobjective Optimisation Criteria in Development of Fuzzy Controllers with Dynamics, Preprints of IFAC Workshop on Control Applications of Optimisation – CAO 2003, Visegrad, pp. 261–266, 2003
34. L.-X. Wang, Adaptive Fuzzy Systems and Control, Prentice-Hall, Englewood Cliffs, NJ, 1994
35. H. Kolmanovsky and N.H. McClamroch, Developments in Nonholonomic Control Ssystems, IEEE Control Systems Magazine, vol. 15, pp. 20–36, 1995
36. H.G. Tanner and K.J. Kyriakopoulos, Backstepping for Nonsmooth Systems, Automatica, vol. 39, pp. 1259–1265, 2003

7

Applications of Fuzzy Logic
in Mobile Robots Control

M. Botros

Department of Computer and Electrical Engineering, Faculty of Engineering,
McMaster University, 1280 Main St. West,
Hamilton, Ontario, Canada L8S 4K1
botrosmw@mcmaster.ca

Fuzzy logic is a mathematical tool that can manipulate human reasoning, concepts and linguistic terms. It suits ill-defined systems since it can handle imprecise information about the system model. In designing controllers for mobile robots, it is very difficult to build a model of the robot environment since the robot tasks usually include dynamic and unstructured environments and noisy sensors.

In this chapter, we present two different approaches for the automatic design of fuzzy inference systems. The first approach is through the use of Neuro-Fuzzy architecture and a learning process to adapt the fuzzy system parameters. The second approach is through the use of Genetic Algorithms (GA) as an optimization tool for selecting the most suitable membership functions and rules for the fuzzy system. The two approaches are illustrated by practical experiments that applied these concepts to the problem of automatic design of fuzzy controllers for mobile robots. A comparison between the two approaches is also presented. Finally, the chapter ends with an application of fuzzy logic to robots communication in multi-robot teams.

7.1 Incorporating Fuzzy Inference Systems and Neural Networks

Fuzzy inference systems have the advantage of simulating the human thinking by using linguistic variables and knowledge base represented by if-then rules. However, they suffer from the difficulty of selecting the best rules and membership functions for controlling a given system. On the other hand, neural networks are parallel structures that can learn and the learning is stored in the synaptic weights of the network. However, this representation of knowledge as synaptic weights can not be acquired by human reasoning. These facts together lead us to thinking of integrating fuzzy inference systems and neural networks to achieve both advantages of simple knowledge representation and similarity to human thinking (fuzzy systems) and the ability to learn (neural networks).

This section will present three different approaches for incorporating fuzzy inference systems and neural networks. After introducing this big picture we will focus on the second approach, "Neuro-Fuzzy systems", and its application in the field of mobile robots control. This approach allows the automatic design of fuzzy system through a learning process that adjusts the parameters of the fuzzy system such as the parameters of the membership functions.

We can distinguish between three different methods in which neural networks can be combined with fuzzy inference systems:

Fuzzy Neurons systems In these systems, the neural network operates on fuzzy inputs and the conventional computational model of the neurons is replaced by a fuzzy model. In the conventional model of the neuron the inputs are first multiplied by the synaptic weights, added then the activation function is applied to the sum. In a typical fuzzy neuron, the membership values of the fuzzy inputs are first calculated (called synaptic operation) then these values are aggregated to produce the membership value of the output [1, 2]. Also the term fuzzy neural networks may be used for neural networks using conventional neuron model while using fuzzy rules in the learning algorithm [3].

Neuro-Fuzzy systems In these systems, a neural network is used to tune the parameters of the fuzzy inference system [8]. These parameters can be the membership functions of the inputs and the outputs, or the weights of the fuzzy rules, or both. Learning algorithms similar to the algorithms used in neural networks are used for iterative tuning of the fuzzy system parameters.

Hybrid Neural Fuzzy systems In the previous type, the role of the neural network was tuning the parameters of the fuzzy system during the training phase but it does not participate during the operation of the fuzzy controller after being trained. However, in this type of systems both the fuzzy and neural networks modules take part in the control of the system [27]. For example, noisy sensory data may be processed by a trained neural network before being applied to the fuzzy inference systems to take control action based on its knowledge base.

The next Sect. (7.2) will present a neuro-fuzzy system architecture along with its learning algorithm and the similarities between this architecture and Radial Basis Function (RBF) Networks. Section (7.3) will show how the principles of neuro-fuzzy systems could be applied to the problem of controlling a mobile robot through an experiment that used neuro-fuzzy controller to enable the robot to avoid obstacles.

7.2 Neuro-Fuzzy Controllers

This section will discuss neuro-fuzzy systems by presenting one widely used architecture for these systems known as ANFIS, which stands for Adaptive Network Based Fuzzy Inference Systems. It presents a network based architecture for fuzzy systems that will enable learning algorithms to adaptively tune the parameters of the fuzzy system.

ANFIS architecture was first introduced by Jang in his original paper with the same name [8]. For the purpose of introducing ANFIS architecture we will consider an example of a fuzzy system with two inputs x_1 and x_2 and one output y. The inference system is Takagi-Sugeno type in which the output is linear function of the inputs. The rule number i can be written as follows:

If $(x_1$ is $A_i)$ and $(x_2$ is $B_i)$ then $(y_i = p_i x_1 + q_i x_2 + r_i)$.

We will use N to denote the number of inputs and R to denote the number of rules. Figure 7.1 shows the structure of an ANFIS system for the fuzzy inference system described above with two inputs $(N = 2)$ and two rules $(R = 2)$. It can be seen that is consists of five layers:

Fuzzification A membership function is defined for every input in each rule. These membership functions are defined using parameters that will be later tuned by the learning algorithm. There is no requirement for the membership function other than being piecewise differentiable. In literature many functions such as Gaussian or triangular are being used. Example of the Gaussian membership function in terms of parameters a_{ij} and b_{ij} is given by:

$$\mu_{ij} = \exp \frac{-(x_j - a_{ij})^2}{2b_{ij}} \quad i = 1, \ldots, R \;\; j = 1, \ldots, N \qquad (7.1)$$

where i is the rule number and j is the input number. The number of neurons in this layer is the product of the number of the inputs and the number of the rules $(N.R.)$.

T-norm In this layer the weight or firing strength of each rule w_i is calculated by applying T-norm operator on the input membership values. Examples of the T-norm functions are Min or Multiplication operators. The number of nodes in this layer is the number of the rules. In the case of the Min operator, w_i will be given as:

$$w_i = \min(\mu_{i1}, \mu_{i2}, \ldots, \mu_{iN}) \qquad (7.2)$$

Similarly, in the case of the Multiplication operator, w_i will be given by:

$$w_i = \prod_{j=1}^{N} \mu_{ij} \qquad (7.3)$$

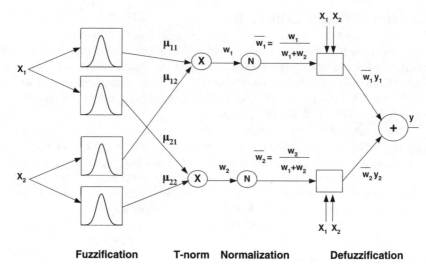

Fig. 7.1. ANFIS architecture for Takagi-Sugeno system with two inputs, one output and two rules

Normalization In this layer the weight of each rule is normalized by the total
    weights of all rules

$$\bar{w}_i = \frac{w_i}{\sum_i w_i} \tag{7.4}$$

Defuzzification The fourth and fifth layers in Fig. 7.1 act as a defuzzification
    operation to obtain a crisp value of the output. In the fourth layer the
    output of each rule is evaluated. For our Takagi-Sugeno example, the
    output is given by:

$$y_i = p_i x_1 + q_i x_2 + r_i \tag{7.5}$$

In the fifth layer, the outputs of different rules are aggregated and the
final output is a weighted sum of the outputs of all rules

$$y = \sum_{i=1}^{R} \bar{w}_i \cdot y_i \tag{7.6}$$

Now, the fuzzy system is defined in terms of:

- Premise parameters: They are the parameters of the membership functions
  $(a_{ij}, b_{ij}$ where $i = 1, \ldots, R$ and $j = 1, \ldots, N)$. Their number is equal to
  $2RN$
- Consequent parameters : They are the parameters of the output rules
  (for example p_i, q_i, r_i where $i = 1, \ldots, R)$. For a single output system,
  the number of these parameters in Takagi-Sugeno case is $R(N + 1)$ since
  we have R rules and in each rule we have $N + 1$ coefficient in the linear
  combination that forms the rule output y_i.

In the following subsection we are going to discuss the learning procedure to tune these parameters but before moving to the learning algorithm we will present a special case of this architecture that will turn to be useful in comparing the performance of Neuro-fuzzy systems with radial basis function (RBF) networks.

Special Case

We now consider a special instance of the previous neuro-fuzzy architecture under the following conditions:

- The membership functions are Gaussian and described by (7.1).
- T-norm is chosen to be the multiplication operator. The weight or firing strength of rule number i is given by:

$$w_i = \mu_{i1}\, \mu_{i2} \cdots \mu_{iN}$$
$$= \exp - \left(\frac{(x_1 - a_{i1})^2}{2b_{i1}} + \cdots + \frac{(x_N - a_{iN})^2}{2b_{iN}} \right) \tag{7.7}$$

- The consequent part of each rule is just a constant (or a singleton), so the output of each rule is just r_i while other terms of the linear combination (such as p_i and q_i) are zeros. Each rule can now be written as:

$$y_i = r_i \quad i = 1, \ldots, R \tag{7.8}$$

Substituting the above equation in (7.6), we can now write the output y as:

$$y = \sum_{i=1}^{R} \bar{w}_i \cdot r_i \tag{7.9}$$

The two equation (7.7) and (7.9) will be later used in comparing this special instance of the neuro-fuzzy architecture with RBF networks.

7.2.1 Learning Algorithm

Now the neuro-fuzzy system is described by a set of premise and consequent parameters and we want to find the best set of parameters that allows system to produce a given desired output y_d for given input vector x. The learning algorithm will use the given input-output pair (x, y_d) to tune these parameter iteratively.

Main learning algorithms for adaptive systems depend on minimizing the square of the error between the output of the system y and the desired output y_d for the same input vector x. Let z be the vector of all parameters to be adjusted:

$$z = [a_{11}, \ldots, a_{RN}, b_{11}, \ldots, b_{RN}, p_1, \ldots, p_R, \ldots, r_1, \ldots, r_R] \tag{7.10}$$

At iteration n, we apply input vector x and we get an output $y(n)$ while the desired output is $y_d(n)$. Our goal is to minimize the square of the error between the actual output and the desired output expressed as the function $v(z(n))$.

$$v(z(n)) = \frac{1}{2}[y(n) - y_d(n)]^2 \tag{7.11}$$

To obtain a learning rule of parameter vector z that implements the gradient descent algorithm, the parameters will be updated such that their change is proportional to the derivative of $v(z)$ with respect to z. The iterative learning rule will be:

$$z(n + 1) = z(n) - \eta \cdot \nabla_z V(z(n)) \tag{7.12}$$

where η is the learning rate. If we express η in terms of another variable, say s, scaled by the norm of $\nabla_z v(z)$ then we have:

$$\eta = \frac{s}{\|\nabla_z v(z(n))\|} \tag{7.13}$$

$$= \frac{s}{\sqrt{\sum_i (\frac{\partial v}{\partial z_i})^2}} \tag{7.14}$$

Combining the last equation and the learning rule, we get

$$z(n + 1) = z(n) - s \cdot \frac{\nabla_z v(z(n))}{\|\nabla_z v(z(n))\|} \tag{7.15}$$

Now, s represents the length of the step used in the direction of the gradient.

As in any adaptive algorithm, there are bounds on η to ensure the convergence of the algorithm. Treatment of this topic can be found in texts such as [10, 11]. Moreover, even if we choose the value to meet this bound, the value of the learning rate still affects the speed of the convergence of the parameters. The small values of η will take more iterations to reach the value of the parameter vector z that minimizes $v(z)$ but it approaches this value with higher accuracy. On the other hand, the large learning rate will result in fast initial reduction of $v(z)$ because it uses larger step size s. However, the value of the parameter vector z will oscillate around the optimal value without approaching it closely since we are changing the parameters by a large increment each time.

The above described effect on the convergence rate can give an idea about how to adjust the learning throughout the learning process. In the initial iterations when we successively get a decrease in the value of $v(z)$, we can increase the step size s (and consequently η) while if the value of $v(z)$ is oscillating between increasing and decreasing, then we are using a large value of learning rate and we should decrease s. A similar heuristic was suggested in [8]. Its main idea is adjusting s throughout the learning process using the following two rules. The value is s can be increased by 10% when coming to four iterations with successive decrease of $v(z)$, while it can be decreased by 10% when two successive iterations of increase and decrease of $v(z)$ are encountered.

Beside the gradient method, other learning algorithms such as least square estimation may be used. Also, a hybrid learning algorithm that uses gradient method and least square estimation was suggested in [8].

7.2.2 Similarity between Neuro-Fuzzy Systems and Radial Basis Function Networks

Radial Basis Function (RBF) Network is a widely used class of neural network models. Its main applications are approximating functions and modeling dynamic systems. In this subsection we will demonstrate how this type of neural network is equivalent to the special case of the neuro-fuzzy system we presented earlier. This equivalence relation shows how neuro fuzzy systems can be used in approximating nonlinear functions and modeling complex dynamic systems as those needed for controlling mobile robots. Another useful result of this relation is applying the learning algorithms of one system to the other.

Figure 7.2 shows a RBF network with two inputs x_1 and x_2. However, we will present the equations describing the network for general input vector $\vec{x}$. The RBF network consists of two main layers:

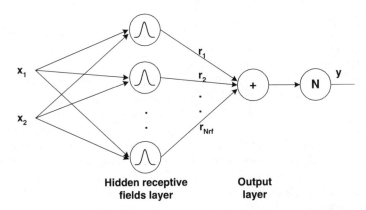

Fig. 7.2. Radial basis function network with two inputs

1. Hidden layer of receptive field units whose number is N_{rf}. The output of each unit w_i is given by:

$$w_i = RF \left(\frac{\|\vec{x} - \vec{a_i}\|}{b_i} \right) \quad i = 1, \ldots, N_{rf} \qquad (7.16)$$

where $\vec{x}$ and $\vec{a_i}$ are vectors of length N, and $RF(.)$ is the receptive field function. If the Gaussian function is used as the receptive field function then w_i will be given by:

$$w_i = \exp\left(\frac{-\|\overrightarrow{x} - \overrightarrow{a_i}\|}{2b_i}\right)$$

$$= \exp-\left(\frac{(x_1 - a_{i1})^2}{2b_i} + \cdots + \frac{(x_N - a_{iN})^2}{2b_i}\right) \qquad (7.17)$$

The vector $\overrightarrow{a_i}$ represents the center of the receptive field number i and b_i represents the radius of the receptive field. The output w_i is large when the input is within radius b_i from the center.

2. Output layer that calculates the output y as a weighted sum of the outputs w_i of the receptive fields.

$$y = \sum_{i=1}^{N_{rf}} w_i \cdot r_i \qquad (7.18)$$

Furthermore, if we consider a scaled version of this output y that is normalized by the the sum of w_i, then y will be given by:

$$y = \sum_{i=1}^{N_{rf}} \frac{w_i}{\sum_i w_i} \cdot r_i$$

$$= \sum_{i=1}^{N_{rf}} \bar{w}_i \cdot r_i \qquad (7.19)$$

We can now explain the function approximating capabilities of the RBF Networks. The domain of the input $\overrightarrow{x}$ is divided into a number of regions equal to N_{rf}. The first layer of the network receives the input vector and outputs the quantities w_i which act as an indicator of the region that contained the input. If $\overrightarrow{x}$ falls in a certain region, then its corresponding w_i will be the most significant one. Next, we can think of the weights r_i as the approximated function value in the region i. In the weighted sum operation, we select which r_i to be the most effective component of y depending on the corresponding value of w_i. This is similar to the case of fuzzy inference system when the consequent part of each rule is $y_i = r_i$. The firing strength of each rule w_i selects which r_i is significant in the output y.

This interesting relation between the RBF Networks and the special case of the neuro-fuzzy systems we pointed earlier was introduced in [9]. It can be shown from comparing the two (7.17) and (7.19) describing the RBF Network and the two (7.7) and (7.9) describing that special case of the neuro-fuzzy system that both equations are equivalent if:

• Number of the rules is equal to the number of the receptive fields $R = N_{rf}$.
• In each rule of the fuzzy system, $b_{i1} = b_{i2} = \cdots = b_{iN}$ in (7.7) and they are equal to b_i of (7.17) in the RBF Network.

The above two conditions shows that the RBF Network is a restrictive case of the neuro-fuzzy system whose rules are only a constant r_i. The neuro-fuzzy system in this case is more general since it allows each input to have different radius b_{ij} in the Gaussian function. This gives some intuition that it can be better in approximating functions compared to RBF Networks since it is more general.

7.3 Experiment 1: Neuro-Fuzzy Controller for Obstacle Avoidance

In this section, we will introduce an experiment that verified the ability of neuro-fuzzy system to approximate nonlinear function using the learning algorithm discussed earlier and then it used the same algorithm to tune the parameters of a neuro-fuzzy controller to enable Khepera robot to navigate safely while avoiding obstacles. This work is performed by Godjevac in [4, 5] and Godjevac and Steele in [6]. We will first present the Khepera robot followed by the structure of the controller then we will present the experiment results and final comments on them.

7.3.1 Khepera Robot

Khepera is a miniature mobile robot that is widely used in laboratories and universities in conducting experiments aiming at developing new control algorithms for autonomous robots. It was developed by the Swiss Federal Institute of Technology and manufactured by K-team [25, 26]. Khepera robot is cylindrical in shape with a diameter of 55 mm and a height of 30 mm. Its weight is about 70 gm. Its small size and weight made it ideal robotic platform for experiments of control algorithms that could be carried out in small environments such as a desktop. The robot is shown in Fig. 7.3.

The robot is supported by two wheels; each wheel is controlled by a DC motor that can rotate in both directions. The variation of the velocities of the two wheels, magnitude and direction, will result in wide variety of resulting trajectories. For example if the two wheels rotate with equal speeds and in same direction, the robot will move in straight line, but if the two velocities are equal in magnitude but different in direction the robot will rotate around its axis.

The robot is equipped with eight infrared sensors. Six of the sensors are distributed on the front side of the robot while the other two are placed on its back. The exact position of the sensors is shown in Fig. 7.4. The same sensor hardware can act as both ambient light intensity sensor and proximity sensor.

To function as proximity sensors, the sensor emits light and receives the reflected light intensity. The measured value is the difference between the received light intensity and the ambient light. This reading has range [0, 1023] and it gives a rough estimate how far the obstacles are. The higher reflected light intensity the closer obstacles are. It should be noted that we cannot find a direct mapping between the sensor reading and the distance from the obstacle, as this reading depends on factors other than the distance to the obstacle

Fig. 7.3. Miniature mobile robot Khepera (with permission of K-team)

Fig. 7.4. The position of the eight sensors on the robot (with permission of K-team)

such as the color of the obstacle. In a similar way, sensors use only receiver part of the device to measure the ambient light intensity and return a value that falls in the range of [0, 1023]. In this experiment the sensors were only used as proximity sensors, but we will present later an experiment involving application of fuzzy logic in multi-robots team that used both functions of the sensors.

7.3.2 The Neuro-Fuzzy Controller

The designed neuro-fuzzy controller had 4 inputs $x_1, \ldots, x_4$ which correspond to the left sensor S_0, the right sensor S_5, the average of the 2 front sensors S_2 and S_3 and the average of the 2 back sensors S_6 and S_7. The locations of this sensors is shown in Fig. 7.4. The outputs of the system are the two speeds of the left and right wheels.

The type of the controller is Takagi-Sugeno with each output rule is a constant (a singleton) instead of linear combination of the inputs in exactly the same way as in the special case of presented in the end of Sect. (7.2). So each rule will be in the form:

If $(x_1$ is $A_i)$ and .. $(x_4$ is $D_i)$, then $(y_{i1} = r_{i1})$ and $(y_{i2} = r_{i2})$

where i is the rule number and the second subscript is used to indicate the number of the output. Each of two output parts $y_{i1} = r_{i1}$ and $y_{i2} = r_{i2}$ is a special case of (7.5) when terms that are proportional to the inputs are zeros.

The number of consequent parameters r_{i1} and r_{i2} for $i = 1, \ldots, R$ is equal to 2R.

The membership functions of the inputs were chosen to be Gaussian with parameters a_{ij}, b_{ij}. They can be given by the equation:

$$\mu_{ij} = \exp -\frac{(x_j - a_{ij})^2}{2b_{ij}} \tag{7.20}$$

where i is the rule index $i = 1, \ldots, R$, and j is the input index $j = 1, \ldots, 4$. The number of these parameters is $2R \cdot N = 8R$.

7.3.3 Experiment Results

The learning algorithm presented in Sect. (7.2.1)was used in adapting the parameters of the above neuro-fuzzy controller. Before being applied to the neuro-fuzzy controller, it was first tested on 3 nonlinear functions and the task was to approximate these functions. The functions are:

$$f_1 = \begin{cases} 0.8 & x_1^2 + x_2^2 > 0.5 \\ 0.2 & \text{otherwise} \end{cases} \tag{7.21}$$

$$f_2 = \sin(3x_1x_2) \tag{7.22}$$

$$f_3 = 4\sin(\pi x_1) + 2\cos(\pi x_2) \tag{7.23}$$

The range of each of x_1, x_2 is $[-1,1]$ and the output range in the three cases is scaled to be in the range $[0,1]$. The three functions are shown in Fig. 7.5. The experiment used neuro-fuzzy system with 7 membership functions for each input for approximating f_1, 5 membership function for each input for approximating f_2, and 3 membership functions in the case of f_3.

The results of the test functions showed that the algorithm was able to tune the parameters of the neuro-fuzzy system such that the square of error between the neuro-fuzzy output and exact function value, $v(z)$, was reduced to value 0.001 after 200 iterations for the second and third test function. Due to the discontinuities in the first function, $v(z)$ in this case remained close to the value of 0.05. For the sake of comparison, the same learning algorithm was used to tune the parameters of three RBF Networks to approximate these functions. The results of the experiments showed that the value of $v(z)$ stayed

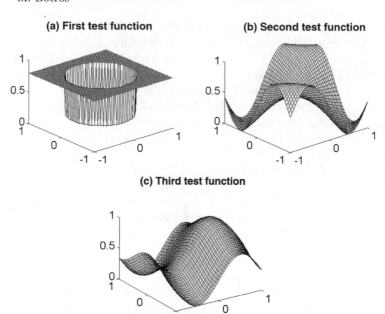

(a) First test function   (b) Second test function

(c) Third test function

Fig. 7.5. Nonlinear test function for the learning algorithm

at a level of 0.1 even after 1000 iterations although the networks had 100 receptive fields for the first and third functions and 80 receptive fields for the second function [5, 6]. In our discussion of the relation between adaptive neuro-fuzzy systems and RBF Networks, we pointed out that first offers more potential for function approximation by providing different radii for different inputs in each rule thus acting as a receptive field but for scaled version of the inputs in such a way that every component of inputs is scaled by a different factor.

Next after this test of the learning algorithm, the same learning procedure was used to tune the neuro-fuzzy controller described in 7.3.2. The controller had 625 rules (R = 625) and the parameters were tuned by input-output examples (sensory information and corresponding output speeds) that represent the behavior to be learned by the robot which is obstacle avoidance in our case. A problem that was encountered after the learning process is that some of the parameters of membership functions were outside the universe of discourse of their corresponding variables. For example, the center of the membership function of input sensor a_{ij} should be within the domain of the input which is [0,1023] in this case. Another problem was that some membership functions overlapped together. The rules containing membership functions that overlapped with others or existed outside the universe of discourse were deleted [4,5]. The performance of the controller after the learning process was tested by placing the robot in an environment with central irregular shape

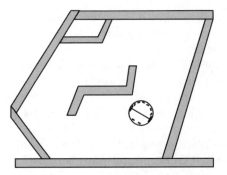

Fig. 7.6. The testing environment

obstacle shown in Fig. 7.6. The robot was successful to wander around this obstacle without colliding with it nor colliding with the walls of the testing environment.

This experiment implemented a successful obstacle avoidance behavior using neuro-fuzzy controller. The two main problems that faced it were the unrestricted membership parameters that could overlap or exist outside the universe of discourse, and the large number of rules that leads to large number of parameter to be learned. A possible solution for the first problem is adding some constraints on the parameters to be learned to keep them inside a desired region. A modified version of the gradient descent, called constrained gradient method, can then be used for learning these parameters taking into account the constraints. One method to reduce the number of parameters is to define fixed membership functions on the universe of discourse of each input variable that will be used in all the rules. This approach was adopted by [7] in designing a controller for the same task but for a different mobile robot. In this case, three membership functions were defined for each input sensor and each membership function was Gaussian with two parameters a and b, so the total number of premise parameters in this case will be $(2N_{mf} . N)$ where N_{mf} is the number of the membership functions of the each input and N is number of inputs as defined through this chapter. On the other hand, the number of premise parameters in the experiment we presented is $(2R . N)$ where R is the number of rules.

7.4 Evolving Fuzzy Inference Systems Using Genetic Algorithms

In the previous sections, we presented our first method for automatic design of fuzzy inference systems. It relied on using neuro-fuzzy architecture and applying a learning procedure for tuning the parameters of the fuzzy system.

In this section, we will present our second method for automatic design of fuzzy inference systems through the use of Genetic Algorithms to evolve the fuzzy system parameters.

Genetic algorithms (GA) are optimization algorithms that mimic the metaphor of natural biological evolution [13]. They are based on Darwinian notation of "Survival of the Fittest". GA operates on a population or a group of individuals representing proposed solutions to the problem. It applies a group of operators on this population to produce a better generation. At each generation, the algorithm selects individuals according to there fitness and uses operators inherited from natural process of evolution to create a generation that is more suited to the problem to be solved. It is clear here the role played by the fitness in the selection of individuals from the previous generation and thus their competition for survival.

The Genetic Algorithms as stated above have 4 basic elements:

- A population of "individuals".
- A notation of "fitness".
- A birth and death cycle controlled by the fitness. The higher the fitness the greater probability the individual has to survive in the next generation.
- A notation of "heritability", where the new generation (offspring) carries some of the properties of the previous generation (parents) through the genetic operators used such as mutation and recombination.

A Genetic Algorithm offers useful properties as an optimization tool. It is applicable to continuous, discrete and mixed optimization problems and it requires no information about the continuity or the differentiability of the function to be optimized [13]. It also can be used for problems of optimization with constraints. The constraints on the parameters to be optimized can be easily translated to constraints on the genetic operators to produce individuals inside the search domain defined by the constraints.

The Genetic algorithm (GA) can be incorporated with fuzzy inference systems to offer a method for optimizing or adjusting the membership functions and learning fuzzy rules. This method solves a main problem in fuzzy systems design which is the absence of a systematic method for choosing the rules and membership functions. The approach of using GA with fuzzy systems is often referred to as evolutionary fuzzy learning [14, 15].

The genetic algorithm can be used to evolve the fuzzy rules of the inference system or the membership functions or both of them. We will present two examples in Sects. (7.5) and (7.6). The first uses GA for learning rules while the other for learning both of the membership functions and rules. The two examples applied the evolutionary fuzzy learning to the problem of controller design of mobile robots. However, each of them represents a different approach for encoding fuzzy rules and genetic operators. Presenting each work will start with brief description of the goal of the experiment, then it will discuss in more details the encoding scheme of the rules, the evolution process, and the resulting robot behavior. The presented methods of encoding the rules and other parameters can be used generally in areas other than robotics with the same concept.

7.5 Experiment 2: Evolving Fuzzy Rules

This section presents an application of genetic algorithm in learning the rule base of fuzzy logic controller. The experiment was performed by F. Hoffmann and G. Pfister [16]. The fuzzy logic controller is Mamdani type and it is required to enable the mobile robot to reach a specified target location while avoiding the obstacles. While the fuzzy rules in this experiment were learned by the genetic algorithm, the fuzzy sets of the input and output variables were designed by hand.

The genetic algorithm used is the messy genetic algorithm (mGA) [20] which is a variation of the conventional genetic algorithm presented in [13]. Some differences exist between them such as the using of cut and slice operator instead of crossover operator. Messy genetic algorithm was chosen because it suits more the variable length chromosome used in this experiment. The following subsections will present the method used to encode the fuzzy rules in the genetic strings, the genetic operators of the messy genetic algorithm, and the evolution process.

7.5.1 Encoding Fuzzy Rules

Each individual fuzzy inference system is represented by a chromosome consisting of many genes. Each rule in the inference system is mapped to a sequence of these genes. To represent the fuzzy rules by a sequence of genes we do the following: First, we assign each input and output a distinct integer label, we also assign labels to fuzzy sets of each variable. For example, we can have five input sensors $d_1, \ldots, d_5$ and an output speed V with fuzzy sets {Very close, Close, Far, Very far} defined for input variables and fuzzy sets {Small, Medium, Large} defined for the output variable. Then, according to the labeling method used, we assign the integers $1, \ldots, 6$ to the variables $d_1, \ldots, d_5, V$. The fuzzy sets of the input variables {Very close, Close, Far, Very far} will be assigned numbers $1, \ldots, 4$ respectively while the fuzzy sets of the output variables {Small, Medium, Large} will be assigned numbers $1, \ldots, 3$ respectively. Second, using this notion we can encode the fuzzy clause (d_1 is Close) to the ordered pair (1,2), by this method we can encode any rule as a sequence of genes each represented as an ordered pair. For example the rule:

$$\text{If} = (d_1 \text{ is far}) = \text{and} = (d_2 \text{ is far}) = \text{then} = (V \text{ is large})$$

will be encoded into the sequence (1, 3) (2, 3) (6, 3). We should note also that the order of the genes (or ordered pairs) inside the rule is not important since input and output variables have different labels. Figure 7.7 shows the general picture of a chromosome

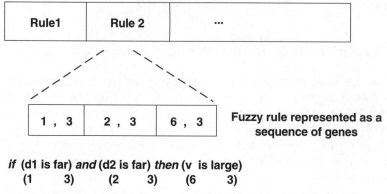

if (d1 is far) **and** (d2 is far) **then** (v is large)
(1   3)   (2   3)   (6   3)

Fig. 7.7. Coding fuzzy rules for messy Genetic Algorithm

Using this encoding scheme will require handling the following situations to resolve any ambiguity:

- An individual is constructed by genetic operators and there was no output clause in the rule. In this case the genetic algorithm needs to randomly add an output clause to rule.
- More than one fuzzy clause may appear in one rule for the same input or output variable. If the repeated variable is input variable, then multiple clauses to the same input can be interpreted as their fuzzy sets being combined in "or" operation. However, if the repeated variable is an output variable then we need only to consider one clause and the others will be neglected.

We note that this method uses variable length chromosome to allow encoding a different number of rules for each individual. Also it does not restrict each rule to have all inputs in the antecedent part thus allowing the evolution of compact set of rules. However, in the next experiment we will present a method of using fixed length chromosome to achieve the same goal of evolving a compact set of rules.

7.5.2 Genetic Operators

The genetic operators used by the mGA are the mutation, and cut and slice operators. The cut and slice operator is applied on the level of the chromosome (on a set of rules) and on single rule level. The general idea of the cut and slice operator is that the two parents are cut at different locations which results in four different segments. Two random segments are randomly selected out of the four segments and concatenated together. This operation bears two differences from the usual crossover operation. First, the positions of cutting in the two parents are different in general so the resulting offspring will have different lengths. Second difference is that it is possible to have the two concatenated strings coming form the same parent.

On the chromosome level, the cut operation splits each parent chromosome into two set of rules while the slice operation concatenates two random sets of rules. Extra completion operation is used here. This operation compares the resulting offspring with the first parent chromosome and missing rules are transferred to the offspring to ensure it is at least as complete as the first parent. The three operations are shown in Fig. 7.8.

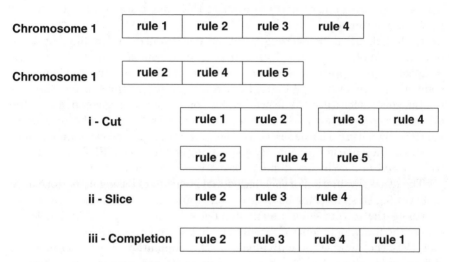

Fig. 7.8. Cut and slice operation on the chromosome level with extra completion operation

The cut and slice operator is also performed, but with lower probability, on a single rule level, that is two parents rules each presented as a sequence of genes are used to produce a new rule. It is desirable here to combine rules that are similar in sense that they have same input variables in there antecedent parts such that the resulting rule will handle an input situation related to the input situation handled by the parent rules. To achieve this, cut and slice operator is preceded by matching process in which the first parent rule is matched to a related rule.

Mutation operator is applied on the rule level or on the clause level. On the clause level the mutation operator can change the integer representing a fuzzy set to another adjacent fuzzy set (Very close to Close for example). While on the rule level, the mutation operator can delete or add a randomly generated input clause.

7.5.3 Evolution Process

As we mentioned earlier, the task of the controller is to enable the mobile robot to reach its target location while avoiding all surrounding obstacles. To achieve this goal, the mobile robot was equipped with 5 ultrasonic sensors for sensing the environment around the robot [16]. These five sensor readings are the input of the fuzzy logic controller (FLC). Since the evolution process takes a lot of time and requires testing each individual FLC in different environments with different obstacles distribution, the evolution process was carried out in a simulated environment and the evolved FLC was tested on the real robot in a real environment.

Each individual in every generation was trained in 30 different environments. Another genetic algorithm is used to generate the training environments. This approach is called co-evolution which involves evolving two competing populations (FLCs and environments populations in our case) simultaneously such that the fitness evaluation of one is at the expense of the other. The co-evolution adds more competition stress to the evolution process which is, by nature, characterized by the competition for survival among individuals of the same generation. By co-evolving environments FLCs are ensured to be tested in new higher difficulty situations every generation.

The fitness evaluation of each individual is calculated using its performance in 30 testing sessions using 30 different environments as mentioned above. Let e_i denote the testing environment number i where $i = 1, \ldots, 30$ and let c_j denote the fuzzy logic controller number j in its population where $j = 1, \ldots, P$ and P is the number of individuals in the population. Each controller c_j will be assigned a value $f(e_i, c_j)$ according to its performance in the testing environment e_i according to the following rules:

- Case 1: Testing session ends when a collision occur. $f(e_i, c_j)$ will be assigned the value $r_c(e_i, c_j)$ which is proportional to the maximum distance from the starting position achieved by the robot.
- Case 2: No collision occurs as in the previous case, but the robot could not reach the specified target location within the allowed number of control steps. $f(e_i, c_j)$ will be assigned the value $r_c(e_i, c_j) + r_d(e_i, c_j)$ where $r_c(e_i, c_j)$ is defined as in previous case and $r_d(e_i, c_j)$ is function of the closest distance from the target achieved by the robot during this session.
- Case 3: The robot reaches its target without collisions. In this case $f(e_i, c_j)$ will be assigned the same previous value plus an extra reward constant value r_t (i.e. $f(e_i, c_j) = r_c(e_i, c_j) + r_d(e_i, c_j) + r_t(e_i, c_j)$).

The performance value of each controller in the given environment e_i is then normalized by scaling each reward type (r_c, r_d and r_t) by the sum of rewards of the same type achieved by other controllers in the same environment. For example, if the controller performance in this environment was assigned a value according to the second case, then the normalized performance value will be given as:

$$\overline{f}(e_i, c_j) = \frac{r_c(e_i, c_j)}{\sum_j r_c(e_i, c_j)} + \frac{r_d(e_i, c_j)}{\sum_j r_d(e_i, c_j)} \qquad (7.24)$$

The fitness value $F(c_j)$ of the controller number j will then be calculated as the sum of its normalized performance values in all the environments:

$$F(c_j) = \sum_{i=1}^{30} \overline{f}(e_i, c_j) \qquad (7.25)$$

The evolution process as described by the genetic operators and fitness evaluation discussed above lasted for enough number of generations to obtain the required behavior of the robot. The evolved controller was then transformed to the real robot and the real behavior is compared with the simulated one. The experiment [16] presented two main results which are the success of the robot to avoid obstacles in real environments, and resemblance between the robot behavior in real and simulated environments. It presented the following two examples as evidence of these results:

- The real environment is a narrow corridor where the robot starts at its beginning and its target position requires the robot to turn after passing the corridor. The robot starts by trying to turn inside the narrow corridor to head directly towards its target but it stops turning to avoid the corridor wall and moves forward for some distance. This behavior was repeated till the robot passed the corridor then turned to reach its target.
- In this case, both the simulated and real environments required the robot to move around an obstacle to reach its target. Robot behaviors in both environments were similar. Same direction of rotation around the object was chosen in both cases.

It was also reported in the experiment that some difficulties faced the robot in real environments with complex obstacle arrangements that were not present in the training simulated environments. Example of these situations is a deadend. The robot was unable to rotate and turn back to exit this deadend.

This example of work in evolving fuzzy logic controller illustrates one method for evolving the rules of the fuzzy inference system. It has chosen the option of using variable length chromosomes to represent possible different number of rules in each individual FLC. The genetic operators have to be tailored to be applied on the chromosome level (set of rules) or single rule level. This is not the only approach. Fuzzy rules could also be evolved using the conventional genetic operators and fixed length chromosome. This approach will be illustrated by the following experiment.

7.6 Experiment 3: Evolving Fuzzy Rules and Membership Functions

This section presents an application of genetic algorithm to learning both the rule base of fuzzy logic controller and the membership functions. Same as the task in the previous work, the controller is required to enable the mobile robot to reach a specified target location while avoiding the obstacles [17]. The genetic algorithm used in this experiment is the conventional genetic algorithm with binary representation and crossover and mutation as genetic operators. The following subsections will present the method used to code the membership functions and fuzzy rules in the genetic strings and the evolution process.

7.6.1 Encoding Membership Functions and Fuzzy Rules

The inputs of the FLC are the 8 infrared sensors of the Khepera robot which was introduced in Sect. 7.3.1. Their reading range is [0,1023] where "0" indicates the absence of an obstacle and "1023" indicates a very close obstacle. For each input variable four triangular membership functions are defined over its universe of discourse [0,1023]. These functions are parametric and expressed in terms of a_1 and a_2 that represent the starting and the ending points of successive membership functions as shown in Fig. 7.9. These parameters are allowed to take one of eight values equally spaced on the range [0,1023] so each of them can be represented by 3 bits, thus a total of 6 bits are required to encode the membership function of each input variable in the binary chromosome representing the FLC. The outputs of the FLC are the two motor speeds of the two wheels of the robot. Their universe of discourse is divided into 4 membership functions with parameters b_1 and b_2 and each of them will be encoded by 3 bits in the binary chromosome.

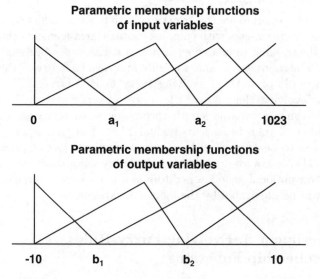

Fig. 7.9. Membership functions for input sensor readings and output speeds

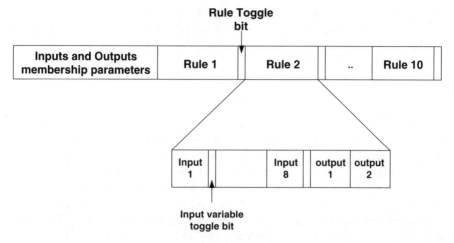

Fig. 7.10. Coding the membership function parameters and fuzzy rules

Beside encoding the parameters of the membership functions of the inputs and the outputs, the chromosome encoded 10 fuzzy rules. Each rule can have any of the 8 input variables in its antecedent part and must have the two output variables in its consequent part. Each input variable will be encoded by three bits, two of them to encode which fuzzy set it belongs to and the last bit is a toggle bit that decides if this input will be included in the rule. The two output variables are encoded in the rule by 2 bits each since no toggle bit is required because each rule must contain the two output variables in the consequent part. Finally, the substring encoding the rule is ended with a rule toggle bit that decides if this rule will be included in the fuzzy inference system or not. Figure 7.10 shows the structure of the binary chromosome that that encodes the inputs, the outputs and the ten rules.

7.6.2 Evolution Process

The evolution process took place in a simulated 2D environment that contained different structure of obstacles beside straight and curved corridors [17]. The evolution process lasted for 200 generations and the task of the robot was to reach the specified target while avoiding the obstacles. The fitness functions used reflected the desired goal of the robot. It had a positive part proportional to the total distance moved by the robot and the number of the check points the robot passed by on its way to the target. Using this evaluation method, individuals that couldn't reach the target but already cut a distance towards the target will be partially rewarded. In addition to this positive part, the fitness function contained a negative part that is proportional to the number of collisions occurred to the robot. Since it is also desired to decrease the number of the rules and number fuzzy sets in each rule to improve the real time performance of the FLC, these factors were also included in the fitness function. Collecting all the above factors we can write the fitness function as follows:

$$f = C_1(distance) + C_2(N_{check\ points}) - C_3(N_{collisions})$$
$$- C_4(N_{fuzzy\ rules}) - C_5(N_{fuzzy\ sets}) \tag{7.26}$$

where $C_1, \ldots, C_5$ are positive constants.

The best fit individual evolved after 200 generations contained 7 rules, 4 of them contained three input variables in their antecedent part, 2 rules with one input variable and one rule with four input variables. This result shows the effect of including the number of fuzzy sets and fuzzy rules in fitness function. The performance of this evolved FLC was tested in two other simulated environments in which it was observed that the robot developed three distinct sub-behaviors which are: passing corridors, wall following and obstacle avoiding. The corridor passing behavior is active when the robot is moving in a narrow path with obstacles on both sides. The wall following behavior become active when the obstacles or walls are sensed on one side of the robot while the obstacle avoidance behavior become active when obstacles are sensed in front of the robot. A relation could be found between each sub-behavior and a subset of the fuzzy rules that support this sub-behavior. The robot switched from one sub-behavior to the other depending on the current situation till its target was reached.

In this experiment, the fitness function included terms to minimize the number of fuzzy rules and fuzzy sets. Limiting the number of the fuzzy rules has the advantage of decreasing the computation time of the fuzzy controller thus improving its real time performance. On the other hand, it can affect the completeness of the knowledge base acquired by the fuzzy system. For example, in the case of fuzzy logic system controlling a robot, insufficient number of rules can result in situations or input sensors combinations that are not covered by the rules.

Next, we will present the result of a comparison performed in [18], it compared the performance of two evolved controllers using GA. The first controller is the one presented in this section and the other one is an evolved neural network controller. The neural network controller consisted of three layers: Input layer with 8 neurons that correspond to the 8 proximity sensors of the Khepera robot, a hidden layer with two neurons and an output layer with two neurons that correspond to the two motor speeds of the robot. The synaptic weights of the neural network were evolved using GA. The purpose of the comparison is investigating possible situations that could not be covered by a fuzzy logic controller with a small number of rules. The best fit individual of the evolved fuzzy logic controller and the best fit individual of the neural network controller were tested in a 2D simulated environment using the robot simulator presented in [19]. The environment consisted of 4 square obstacles with sharp corners which could be hard for the robot to detect if it is heading

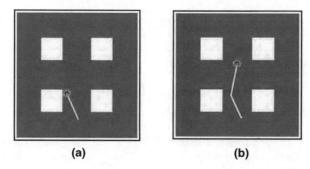

(a)   **(b)**

Fig. 7.11. Comparison of the behaviors of the evolved FLC (*left*) and neural network controller (*right*)

towards the corner with an angle such that only one of its sensor is indicating the presence of an obstacle.

Results of the robot behavior in the fuzzy logic controller case (Fig. 7.11(a)) showed that this situation was an example of input sensor combination that was not covered by the fuzzy rules and the robot could not avoid the obstacle. In the evolved neural network case (Fig. 7.11(b)), the left sensor reading, which was the most active when the robot was heading towards the corner, was able to drive the neural network output in a way to turn the robot before colliding with the corner. A suggested solution to this problem can be made by setting the relative importance of the "obstacle avoidance" and "number of fuzzy rules" parts in the fitness by changing the constants C_3 and C_4 in the fitness function. In the experiment, C_3 which is the coefficient of "obstacle avoidance" part was 3 while C_4 was 100. By changing these two values, the fitness evaluation can emphasize the importance of "obstacle avoidance" more than evolving a controller with a small number of rules.

The previous two experiments were examples of different approaches of evolving FLCs which are evolving the fuzzy rules and evolving the fuzzy system parameters along with the rules. The second approach is more automated one for designing FLCs and reduces any initial choices made by the human designer in choosing the membership functions. The experiments used two different encoding schemes for the fuzzy rules. The first was using variable length chromosome to evolve FLC with no restriction on the rules number or their inputs number, while the other encoding scheme used a fixed length chromosome to simplify the genetic operators while using a rule toggle variables to limit the number of rules included in the inference systems. The small number of rules and any other desired property in the FLC could be included in the fitness function to credit individuals that attain these properties. Table 7.1 shows a comparison between these two different approaches in evolving fuzzy logic controllers.

Table 7.1. Comparison between the two presented approaches for evolving fuzzy logic controllers

	Experiment in Sect. 7.5	Experiment in Sect. 7.6
What is evolved	– Fuzzy rules	– Fuzzy rules and membership functions
Coding Method	– Variable length chromosome – Integer encoding – No max. number of rules	– Fixed length chromosome – Binary encoding – Max. no. of rules is specified.
Fitness evaluation	– Performance of robot	– Performance of robot and parameters of FLC (no. of rules)

7.7 Comparison of GA and Neuro-Fuzzy Approach for Automatic Design of Fuzzy Systems

We have presented two different approaches for the automatic design of fuzzy inference systems. The first approach uses neuro-fuzzy architecture and the other uses GA to evolve fuzzy systems. Each approach has its own learning method. The neuro-fuzzy architecture uses supervised learning where the parameters of the fuzzy system are tuned using given input-output examples that represent the situations or behaviors to be learned by the robot. The GA approach works by specifying a fitness function and selecting individual fuzzy systems with high fitness for reproduction. By this process, it learns the required behaviors that imply a high fitness value. It does not provide input-output examples for the desired behaviors but these behaviors are learned through successive generations and fitness evaluation in different environments that are either real or simulated. This method adds more flexibility in specifying the required behaviors through designing environments with different degrees of difficulty.

Also, each approach handles the restriction on the range of membership parameters in different ways. Neuro-fuzzy approach can be modified such that the constraints representing the feasible range of fuzzy system parameters are included in the learning rule. One of the possible learning procedures for including these restrictions is the constrained gradient descent method. In the GA approach, constraints on the parameters can be translated to genetic operators that result in production of individuals with parameters within the feasible range. Another method for restricting the parameters in the feasible range is through fitness evaluation. Individuals with parameters outside the desired range are assigned low fitness and consequently less chance of reproduction. As the time of evolution advances, generations will contain fewer individuals with these undesirable parameters.

Another point in favor of the GA approach is its way of handling the number of rules of the fuzzy system. The number of rules can be variable by using variable length chromosome or we can specify maximum number of rules and include information in each chromosome indicating which rules to be used by the fuzzy controller. On the other hand, neuro-fuzzy approach needs to specify the number of rules prior to the learning procedure. If the performance of resulting fuzzy system after learning is sot satisfactory, the number of rules will be changed and the learning process will be repeated.

7.8 Application of Fuzzy Logic in Multi-Robot Communication

The goal of employing a team of cooperative multi-robots is achieving a task that a single robot can not achieve by itself. Potential advantages of using multi-robots teams include increasing the reliability of the system, fault tolerance, simpler robot design and exploring new areas of applications [22]. On the other hand, several issues arise in the control of multi-robots. One of these issues is the design of the communication between robots to exchange possible information about their sensor readings, locations and goals. Communication among robots will affect their cooperative interaction and will consequently affect the performance of the whole system.

In the next section we will present an experiment that involves application of fuzzy logic in the field of multi-robot communication. The work in this experiment suggests replacing the communication of sensory information in their crisp form by a communication protocol that uses linguistic variables. It applies this new communication protocol to a team of two robots that cooperatively push an obstacle [21].

7.9 Experiment 4: Communication in Robots Team Using Fuzzy Linguistic Variables

This experiment was performed by Molina et al. [21]. It presents a method of communication among robots in the same team that depends on sharing sensory information in their fuzzy form instead of their crisp value. The task required from the robots is to align themselves to a certain obstacle and to push it cooperatively. Figure 7.12 shows the environment of the experiment and the object to be pushed. The object is designed to have two light sources inside it and the robots will use their light sensors to align themselves in front of these two light sources.

The team of robots consists of two individuals of Khepera robot that was described in Sect. (7.3.1). Each robot has 8 sensors that are arranged on the robot as shown in Fig. 7.4. Each sensor can act as ambient light sensor and proximity sensor. In either cases, the range of the reading is [0, 1023]. However the values of the sensor reading are inaccurate and depend on many factors

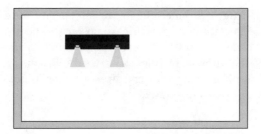

Fig. 7.12. The environment of the experiment

such as the color and material of the obstacle. This uncertain property of the sensors led to considering the idea of sharing the sensory information in their fuzzy form instead of their crisp value.

The task of pushing the obstacle by the two robots is divided into two main tasks that require communication between the robots:

- The first robot will find the object will align itself in front of the obstacle with a short distance between them.
- The first robot will send its alignment information to the other robot. The second robot will use this information to align itself in the same way with respect to the obstacle.

Finally, when these two tasks are performed, the two robots will start pushing the obstacle together. The experiment used an obstacle with 2 lights inside it such that alignment information will be in terms of the readings of both the proximity and light sensors. Communicating the proximity sensors information will ensure that the two robots are at same distance for the obstacle while the communicating the light sensors information will ensure that the two robots have the same orientation relative to the light sources in the object.

To achieve the first task, the first robot has to find the obstacle. This can be achieved by letting the robot to "wander" while using its proximity sensors to take the direction which agrees with the sensor indicating highest reading, i.e. "obstacle approaching". As it can be seen, this is the opposite of the obstacle avoidance behavior discussed in earlier experiments in this chapter. The simplest form of controller to achieve this task can be a reactive controller in which the output motor speeds are direct function of, or possibly a linear combination of, the input sensor readings. This approach was chosen by the experiment designers to let their effort be focused on the second task which includes the communication between the robots. Series of reactive controllers for similar and other tasks are introduced in [23] while other possible controllers for this task can be found in [24]. Now, when the first robot finds the object, it has to be aligned to a certain position with respect to it which is in front of the first light source (see Fig. 7.13(b)). The robot uses fuzzy variables associated with the proximity and light sensors to get to this position. For

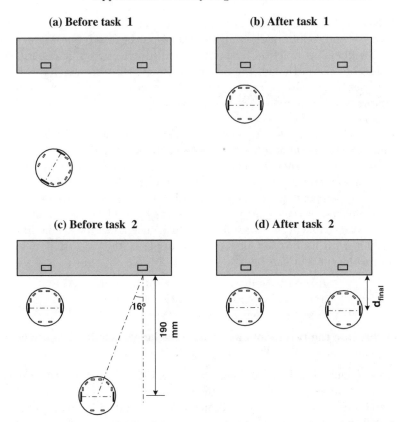

Fig. 7.13. Results of the experiment showing the position of robots before and after each task

example, when the two front light sensors are "High" and proximity sensors are "Near", the first robot is considered aligned and thus it starts communication with the second robot.

The second task starts when the first robot is aligned to the obstacle and with a distance 7 mm between them. The second robot starts from a distance of 19 cm but making an angle to its required final alignment as shown in Fig. 7.13c. The communication is then allowed to start between the two robots. The experiment investigated three forms of communicating the fuzzy form of this sensor information:

- The two robots have the same fuzzy sets for the sensor data, for example {Near, Medium, Far}, and each set has the same membership function. In this case, the first robot sends the fuzzy label of the sensor reading along with its membership values, for example it can send: Near (0.7), Medium (0.3), Far (0).

- The second case is similar to the first one, but the membership functions of the second robot are different from those in the first robot. For example, the membership function medium is defined over two different ranges in each robot. Even though the robots are same, this case simulates a practical case in which the robot team is heterogeneous and each member is equipped with different sensors and different membership functions for the sensors information.

- The third case considers reducing the information communicated to a minimum level. The first robot tries to send to the second one an "idea" or "concept" about the current situation without sending each sensor reading. For implementing this situation, the two robots had the same fuzzy sets and membership functions for each sensor. The first robot processes the membership values of the sensor inputs by a set of rules to obtain the "concept" that will be transmitted. This concept will be a new linguistic variable with its own set of labels. Example of the rules that is used to obtain the concept can be:

> If   (Proximity sensor 3 activation Near > 0.7)
> and   (Light sensor 2 activation Low < 0.3)
> then (Transmit concept $=$ Near)

In this case the two robots have the same labels definition of the concept and same set of rules that calculates the concept.

Experiments in real environments were carried and the three types of communications were tested. The experiment was repeated 50 times for each type of communication and the average of the final distance between the second robot and the obstacle d_{final} was calculated for each case. The results showed that the three methods of communicating the fuzzified sensory information enabled the second robot to a come close to the obstacle with an average distance approximately equal to that of the first robot (about 7.5–8 mm) as shown in Fig. 7.13-d. The average final distances of the three methods were 7.56, 8.09 and 7.15 mm respectively [21]. It is noted that the largest distance occurred in the second method because of the difference in the definition of the membership functions in both robots.

To compare the results with the communication methods based on the crisp values of the sensory information the same experiment was repeated again and the second robot was given a time of 5 minutes to align itself in position such that its sensor readings match those of the first robot. In all 50 trials of the experiment, the second robot was unable to align itself to the object, since no one-to-one relationship can be found between the sensor reading and the distance to the object because of the noisy nature of the sensors.

This experiment [21] proposed communication methods that rely on linguistic variables. The results suggest that they are more robust than sharing the information in their crisp form. It also emphasizes the ability of fuzzy logic to handle noise in the input variables of the system.

7.10 Summary

Fuzzy logic is characterized by the ability to handle human knowledge and reasoning through linguistic variables and a knowledge base represented by if-then rules. However it lacks a systematic method for designing membership functions and rules. Neuro-fuzzy systems and genetic algorithms were presented as two possible approaches for automatic design of fuzzy inference systems.

In the neuro-fuzzy approach, designing the parameters of the fuzzy system is achieved through a network-like structure of the fuzzy system and a learning algorithm such as the gradient descent method. An experiment that aimed at teaching the robot avoiding obstacles was introduced. The issues of decreasing the size of the learned parameters, and restricting the range of membership function parameters were discussed.

In the genetic algorithm approach, evolution is used to learn the rules and the membership functions. Two experiments whose goal was teaching the robot obstacle avoidance and target seeking were presented. Through the two experiments different possible issues of evolving fuzzy systems were discussed. These issues include possible encoding schemes, genetic operators and fitness functions. A critical issue of the number of rules was considered by two different methods in each experiment. Variable length chromosome was adopted in the first experiment to allow different individuals to have different numbers of rules. On the other hand, the second method uses fixed length chromosome that implies a maximum number of rules while including the number of rules in the fitness function.

Also, fuzzy logic was shown to be useful in the mobile robots communication. An experiment was presented in which the goal of the communication between robots is to align themselves to certain obstacle and push it cooperatively. The results of the presented experiment showed how the communication based on the fuzzy variables was successful in achieving the task more than the communication based on the crisp values of the sensor readings.

Acknowledgement

The author would like to thank S. Kirolos and S. Mercorious for their continuing support.

References

1. J. Godjevac, "Comparison between Classical and Fuzzy Neurons," EUFIT, Vol. 3, pp. 1326–1330, Aachen, Germany, 1994.
2. H. Narazaki and A. Ralescu, "A Synthesis Method for Multi-Layered Neural Network using Fuzzy Sets," In IJCAI-91 Workshop on Fuzzy Logic in Artificial Intelligence, pp. 54–66, Sydney, 1991.

3. S. Halgamuge and M. Glenser, "Neural Networks in Designing Fuzzy Systems for Real World Applications," Fuzzy Sets and Systems, 65:1–12, 1994.
4. J. Godjevac, "Comparative Study of Fuzzy Control, Neural Network Control and Neuro-Fuzzy Control," In Fuzzy Set Theory and Advanced Mathematical Applications, D. Ruan Ed., Kluwer Academic, Chap. 12, pp. 291–322, 1995.
5. J. Godjevac, "A Learning Procedure for a Fuzzy System: Application to Obstacle Avoidance," International Symposium on Fuzzy Logic, pp. 142–148, Zurich, Switzerland, 1995.
6. J. Godjevac and N. Steele, "Adaptive Neuro-Fuzzy Controller for Navigation of Mobile Robot," Fifth IEEE International Conference on Fuzzy Systems, Vol. 1, pp. 136–142, 1996.
7. X. Wang and S. Yang, "A Neuro-Fuzzy Approach to Obstacle Avoidance of a Nonholonomic Mobile Robot," Proceedings of Advanced Intelligent Mechatronics, IEEE/ASME, 2003.
8. J. Jang, "ANFIS: Adaptive-Network-Based Fuzzy Inference System," IEEE Transactions on Systems, Man and Cybernetics, Vol. 23, No. 3, pp. 665–685, 1993.
9. J. Jang and C. Sun, "Functional Equivalence Between Radial Basis Function Networks and Fuzzy Inference Systems," IEEE Trans. Neural Networks, Vol. 4, No. 1, pp. 156–159, 1993.
10. S. Haykin, "Adaptive Filter Theory," Prentence Hall, New Jersey, 1986.
11. B. Widrow and S. Stearns, "Adaptive Filter Processing," Prentence Hall, New Jersey, 1986.
12. H. Nomura, I. Hayahi, and N. Wakami, "A Learning Method of Fuzzy Inference Rules by Descent Method," In Proceedings of IEEE Int. Conf. on Fuzzy Systems, pp. 203–210, San Diego, 1992.
13. D. Goldberg, "Genetic Algorithms in Search, Optimization and Machine Learning," Addison Wesley Publishing Company, 1989.
14. Y. Shi, R. Eberhart and Y. Chen, "Implemantation of Evolutionary Fuzzy Systems," IEEE Trans. on Fuzzy Systems, Vol. 7, No. 2, pp. 109–119, 1999.
15. M. Russo, "Genetic Fuzzy Learning," IEEE Trans. on Evolutionary Computation, Vol. 4, No. 3, pp. 259–273, 2000.
16. F. Hoffmann and G. Pfister, "Evolutionary Design of a Fuzzy Knowledge Base for a Mobile Robot," Int. Journal of Approximate Reasoning, Vol. 17, pp. 447–469, 1997.
17. S. Lee and S. Cho, "Emergent Behaviors of a Fuzzy Sensory-Motor Controller Evolved by Genetic Algorithm," IEEE Transaction Systems, Man and Cybernetics (B), Vol. 31, No. 6, pp. 919–929, 2001.
18. M. Botros "Evolving Neural Network Based Controllers for Autonomous Robots Using Genetic Algorithms," Master Thesis, Cairo University, Egypt, 2003.
19. T. Nilson, "Kiks: Kiks is a Khepera Simulator," Master Thesis, Computer Science Department, Umeå University, Sweden, March 2001.
20. D. Goldberg, B. Krob and K. Deb, "Messy Gentic Algorithms Motivations, Analysis and First Results," Complex Systems, Vol. 3, pp. 493–530, 1989.
21. J. Molina, V. Matellan and L. Sommaruga, "Fuzzy Multi-agent Interaction," IEEE International Conference on Systems, Man and Cybernetics, Vol. 3, pp. 1950–1955, 1996.
22. R. Arkin, "Behavior-Based Robotics," The MIT Press, Cambridge, 1998.
23. V. Braitenberg, "Vehicles: Experiments in Synthetic Psychology," The MIT Press, Cambridge, 1984.

24. M. Botros, "Evolving Controllers for Miniature Robot," In Evolvable Machines: Theory & Practice, Chap. 10, N. Nedjah and L. de Macedo Mourelle, Editors, Kluwer Academic, 2004, In Press.
25. K-Team, "Khepera User Manual," Lasuanne, Switzerland, 1999.
26. F. Mondada, F. Franz and I. Paolo, "Mobile Robot Miniaturisation: A Tool for Investigation in Control Algorithm," Proceedings of the Third International Symposium on Experimental Robotics, Kyoto, Japan, 1993.
27. Wei Li, "A Hybrid Neuro-Fuzzy System for Sensor Based Robot Navigation in Unknown Environments," Proceedings of the American Control Conference, Vol. 4, pp. 2749–2753, 1995.

8

Modeling the Tennessee Eastman Chemical Process Reactor Using Fuzzy Logic

A.F. Sheta

Computers and Systems Department, The Electronics Research Institute (ERI), El-Tahrir Street, Dokky, Giza, Egypt
`asheta1@eri.sci.eg`

The goal of this research is to introduce the challenges associated with modeling nonlinear dynamical systems. Some of these challenges are the nonlinear relationships among system variables, the presence of uncertainty and noise in the measurements. In this research, we have three goals to achieve: (1) provide a review of system identification using Least Square Estimation (LSE); (2) give an introduction to the Takagi-Sugeno (TS) algorithm for developing fuzzy model structure; (3) build a fuzzy model for the Tennessee Eastman (TE) chemical reactor under various operating conditions; The LSE method and the fuzzy method, were used to solve the modeling problem for the TE chemical process reactor. The developed results show that fuzzy models can efficiently perform like the actual chemical reactor and with a high modeling capabilities.

8.1 Introduction

It is important to be able to develop models for real world applications for dynamical systems. These models need to simulate actual system behavior in cases where limited a priori knowledge about systems structure exists. Identification of nonlinear systems is considered a difficult problem. The reason is identifying a structure for nonlinear systems usually involves two major steps: (1) the selection of a model structure with a certain set of parameters and (2) the selection of an algorithm to estimate these parameters. The later issue usually biases the former one.

The use of mathematical models is highly required in all fields of engineering. In reality, a major part of the engineering research handle the idea of developing a good design based on mathematical models. Modeling and identification of nonlinear systems is an application dependent problem. In many real-world applications, it is proposed to use models that describe the relationships between system variables in terms of a mathematical formula like difference or differential equations [1–3].

To build a system model, we need to know how its variables are inter-related. Therefore, we can call such a relationship between observed inputs a system model. Models can have various structures and be shaped with varying degree of mathematical format. The intended use of the model specifies the degree of details needed [4, 5]. A dynamic system can be described by two types of models: input-output models [6] and state-space models [7]. In the following section, we describe the input-output model, which is adopted in this research.

8.1.1 Input-Output Models

An input-output model describes a dynamic system based on input and output data. In the discrete-time domain, an input-output model can be of the Auto-Regressive Moving Average (ARMA) type or the parametric Wiener or Hammerstein type model [8,9].

An input-output model assumes that the system output can be predicted by the past inputs and outputs of the system. If the system is further supposed to be deterministic, time invariant, single-input single-output (SISO), the input-output model becomes:

$$y(k) = f(y(k-1), y(k-2), \ldots, y(k-n),$$
$$u(k-1), u(k-2), \ldots, u(k-m)) \qquad (8.1)$$

$u(k)$ and $y(k)$ represent the input-output pairs of the system at time k. n and m are the number of past outputs and the number of past inputs, respectively. n is also called the order of the system. In practice m is usually smaller than or equal to n. Depending on the function f, the system equation can be described as linear, (8.2), (8.3), (8.4), or nonlinear, (8.5), (8.6).

For linear systems, f is a linear function. The system equation can be written as:

$$A(q^{-1})y(k) = B(q^{-1})u(k) \qquad (8.2)$$

The two polynomials $A(q^{-1})$ and $B(q^{-1})$ can be expanded to form the system equation as:

$$y(k) + a_1 y(k-1) + \cdots + a_n y(k-n) = b_1 u(k-1) + \cdots + b_m u(k-m) \qquad (8.3)$$

Thus:

$$y(k) = b_1 u(k-1) + b_2 u(k-2) + \cdots + b_m u(k-m)$$
$$- a_1 y(k-1) - a_2 y(k-2) - \cdots - a_n y(k-n) \qquad (8.4)$$

$a_i(i = 1, 2, \ldots, n)$ and $b_i(i = 1, 2, \ldots, m)$ are real constants. $u(k), y(k)$ represents the input-output pairs of the system at time k.

For nonlinear systems, f is a nonlinear function. f can be a static nonlinear function which maps the past inputs and outputs to a new output. An example of a nonlinear system is the simple Hammerstein model. It is a series connection of a static nonlinear and a linear transfer function. The equation for calculating the output signal $y(k)$ from a given input signal $u(k)$ is given as:

$$v(k) = r_0 + r_1 u(k) + r_2 u^2(k) + \cdots + r_p u^p(k)$$
$$A(q^{-1})y(k) = B(q^{-1})v(k) \tag{8.5}$$

where:

$$A(q^{-1}) = 1 + a_1 + \cdots + a_m q^{-n}$$
$$B(q^{-1}) = b_1 q^{-1} + \cdots + b_m q^{-m} \tag{8.6}$$

The simple Hammerstein model is one of the most widely-known nonlinear process models.

8.2 Traditional Modeling

Least-Square Estimation (LSE) used to solve the parameter estimation problem for both linear and nonlinear systems. To solve the parameter estimation problem for the linear system given in (8.4), we need to collect a data set of the system input and output u and y, respectively, and build the regression matrix ϕ as given in (8.7).

$$\phi =$$
$$\begin{pmatrix} u(k-1) & \cdots & u(k-1-m) & -y(k-1) & \cdots & -y(k-n) \\ u(k) & \cdots & u(k-m) & -y(k) & \cdots & -y(k-n+1) \\ \vdots & & \vdots & \vdots & & \vdots \\ u(k-1+N) & \cdots & u(k-1-m+N) & -y(k-1+N) & \cdots & -y(k-n+N) \end{pmatrix}$$
$$\tag{8.7}$$

where:

$$\theta = \begin{pmatrix} b_1 \\ \vdots \\ b_m \\ a_1 \\ \vdots \\ a_n \end{pmatrix} \tag{8.8}$$

and

$$\hat{y} = \begin{pmatrix} y(k) \\ y(k+1) \\ \vdots \\ y(k+N) \end{pmatrix} \tag{8.9}$$

Thus, we can compute the estimated response of $\hat{y}$ as follows:

$$\hat{y} = \phi\theta \tag{8.10}$$

N is the number of samples which is chosen sufficiently large. To estimate the best set of parameters $\hat{\theta}$, we have to minimize the error difference between the observed and estimated outputs y and $\hat{y}$, respectively. The least square minimization criteria can be written as:

$$E(\theta, k) = \frac{1}{2} \sum_{k=1}^{N} (y(k) - \hat{y}(k))^2 \tag{8.11}$$

The least squares solution yields the normal equation:

$$(\phi^T\phi)\theta = \phi^T y \tag{8.12}$$

which has the solution:

$$\hat{\theta} = (\phi^T\phi)^{-1}\phi^T y \tag{8.13}$$

Thus, from (8.13), we can compute the model parameters which represent the solution of our problem. The best set of parameters are reached when $E(\theta, k)$ is minimal.

8.3 Fuzzy Logic (FL) Modeling

Fuzzy modeling has been increasingly recognized as a powerful paradigm [10–12]. In the past few decades, many approaches to fuzzy modeling proposed to solve variety of modeling and control problems. These approaches include fuzzy relational modeling [13], fuzzy linear models which is also called Sugeno-Takagi models [14] and qualitative linguistic models [15]. Fuzzy models have the capabilities to approximate any smooth function as stated in [6,16]. Many classes of nonlinear systems were modeled using the Takagi-Sugeno (TS) fuzzy models [17,18].

The work developed in [19] was successfully applied to solve the modeling and control problem for multi-input single output (MISO) system process [12,20]. This is why we adopt the use of fuzzy logic to build a dynamic model for the Tennessee Eastman Chemical process reactor.

8.3.1 Fuzzy Identification Approach

Recently, modeling the dynamics of complex industrial processes using artificial intelligence and soft-computing techniques attracted a considerable attention [21]. These techniques include neural networks [22], fuzzy logic [17], genetic algorithms [23] and genetic programming [24, 25]. The structure of the model (i.e. the model order) can be determined by the user according to his/her prior knowledge about the process and/or by comparing various selected model structure in terms of the error minimization between the observed system response and the output estimated by the developed model.

Fuzzy modeling and identification of nonlinear systems from data collected for nonlinear processes attracted a large attention in the past decades [10–12]. The methodology to build a fuzzy model for nonlinear dynamical system can be summarized in the following steps:

- Use the set of measurements $u(k)$, $y(k)$ and the user defined parameters σ to find $\hat{y}(k)$.
- Compute the antecedent membership function from the cluster parameters.
- Estimate the consequence parameters using LSE.

We can write the relationship between the system inputs and outputs as follows:

$$y(k+1) = f(y(k), u(k)) \tag{8.14}$$

where $u(k)$ and $y(k)$ are the input and output at time k and f is a static nonlinear function.

One of the most common models for modeling nonlinear systems is the Nonlinear Auto-Regressive with eXogenous input (NARX) model which can be represented as follows:

$$\begin{aligned} y(k+1) = f(y(k), y(k-1), \ldots, y(k-n+1), \\ u(k), u(k-1), \ldots, u(k-m+1)) \end{aligned} \tag{8.15}$$

where $u(k), \ldots, u(k-m+1)$ and $y(k), \ldots, y(k-n+1)$ denote the past model inputs and outputs, respectively. n and m are integer related to the model order. A singleton fuzzy model of a dynamical system can be described by a set of rules as follows:

$$\begin{aligned} R_i : \ &\textbf{If } y(k) \textbf{ is } A_1 \textbf{ and } \ldots \textbf{ and } y(k-n+1) \textbf{ is } A_n \\ &\textbf{and } u(k) \textbf{ is } B_1 \textbf{ and } \ldots \textbf{ and } u(k-m+1) \textbf{ is } B_m \\ &\textbf{then } y(k+1) \textbf{ is } c_i \end{aligned} \tag{8.16}$$

Using the fuzzy modeling technique, we compute the values of the parameters for the fuzzy model. These parameters include: (1) antecedent membership functions and their consequence polynomials; (2) the number of rules (clusters) σ; The user may try different values for σ till the best modeling performance and the minimal modeling error are reached.

8.3.2 Fuzzy Clustering

Given the regression matrix ϕ and the specified number of clusters σ, Gustanfson-Kessel (GK) algorithm [26] can be applied. This algorithm can be summarized as follows:

- Compute the fuzzy partition matrix $U = [\gamma_{ik}]_{\sigma \times N}$ with $\gamma_{ik} \in [0,1]$. i stand for the rule number.
- Compute the prototype matrix, $V = [v_1, \ldots, v_\sigma]$
- Compute the set of cluster covariance matrices $S = [S_1, \ldots, S_\sigma]$. S_i are positive definite matrices in $R^{(p+1) \times (p+1)}$. p is the dimension of the antecedent space.

Given the triple, (U, V, F) the antecedent membership functions and the consequence parameters A_i, B_i and c_i can be computed.

8.3.3 Gustanfson-Kessel (GK) Algorithm

Consider z, the input output data matrix, where $z = [u, y]$, the number of clusters σ and some $\epsilon > 0$ are given. A detailed description of the Gustanfson-Kessel (GK) algorithm for a multi-input multi-output (MIMO) system can be found in [27].

The goal of this algorithm is to compute the cluster means and the covariance matrix. The description of the Gustanfson-Kessel (GK) algorithm described in [17] is given below with some change in notation.

8.3.4 Estimation of the Consequent Parameters

Let Φ be the matrix $[\phi, 1]$ and the matrix W_i be a diagonal matrix having a membership degree γ_{ik} as its kth diagonal element. Assuming that the columns of the matrix Φ are linearly independent and $\gamma_{ik} > 0$, thus:

$$\hat{\theta} = (\Phi^T W_i \Phi)^{-1} \Phi^T W_i y \qquad (8.17)$$

$\hat{\theta}$ contains the best values of the model parameters computed using least-square estimation knowing that $y = \Phi^T \theta + \epsilon$

8.4 Tennessee Eastman Chemical Process

The Tennessee Eastman process is considered a standard industrial process that is used as a test-bed for the purpose of developing, studying and evaluating new technologies for process control and system identification [28]. The Tennessee Eastman process consists of a number of chemical subsystems. They include, the reactor, the separator, the recycle arrangement and others. The detailed description of the Tennessee Eastman Chemical Process can be found in [29, 30].

Algorithm 8.1 Gustanfson-Kessel (GK) Algorithm

---

1. **Repeat for** $j = 1, 2, \ldots$
2. **step1: Compute cluster means:**

$$v_i^j = \frac{\sum_{k=1}^{N} (\gamma_{ik}^{j-1})^m z_k}{\sum_{k=1}^{N} (\gamma_{ik}^{j-1})^m}$$

3. **step2: Compute covariance matrices:**

$$S_i = \frac{\sum_{k=1}^{N} (\gamma_{ik}^{j-1})^m (z_k - v_i^j)(z_k - v_i^j)^T}{\sum_{k=1}^{N} (\gamma_{ik}^{j-1})^m}$$

5. **step3: Compute distances:**

$$d^2(z, v_i^j) = (z_k - v_i^j)^T \left(det(S_i)^{\frac{1}{p+1}} S_i^{-1} \right) (z_k - v_i^j)$$

6. **step 4: Update partition matrix:**

$$\text{if} \quad d^2(z_k, v_i^j) > 0 \quad \text{for} \quad 1 \le i \le \sigma \quad \text{and} \quad 1 \le k \le N,$$

$$\gamma_{ik}^j = \frac{1}{\sum_{j=1}^{\sigma} (d(z_k, v_i^j)/d(z_k, v_i^j))^{2/(m-1)}}$$

otherwise

$$\gamma_{ik}^j = 0 \quad \text{if} \quad d^2\left(z_k, v_i^j\right) > 0 \quad \text{and} \quad \gamma_{ik}^j \in [0, 1]$$

$$\text{with} \quad \sum_{i=1}^{\sigma} \gamma_{ik}^j = 1$$

until $\|U^j - U^{j-1}\| < \epsilon$

8.4.1 Data Set

During our exploration we have investigated the possibility of finding a data set that represents the process under study. The data used in this article was downloaded from [31]. The data was used by author in [29, 30, 32] to test variety of system identification and control techniques. The data set was collected under various operating conditions [32].

Figures 8.1, 8.2, 8.3, 8.4 show the characteristics of the data sets collected to model the reactor level, reactor pressure, reactor coolant temperature and the reactor temperature.

8.4.2 Case Study: The Reactor

In our case study, we focused on the problem of modeling the Tennessee Eastman Chemical reactor. The modeling process of the reactor can be divided

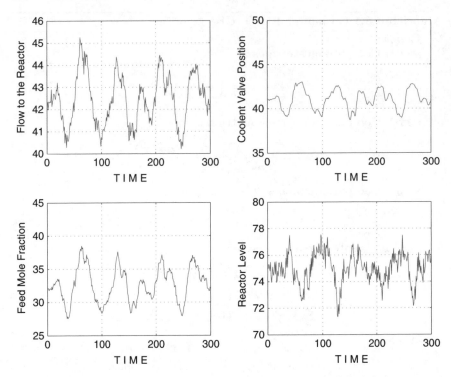

Fig. 8.1. The Input and Output Data for The Reactor Level

into four sub-problems. Each sub-problem has four input variables and has a single output variables.

For the process under study, we have four main inputs $u_1(k),u_2(k),u_3(k),$ $y(k-1)$ and the output $y(k)$. $u_1(k), u_2(k), u_3(k)$ stand for the Flow to the Reactor, Coolant Valve Position and the Feed Mole Fraction. The output $y(k)$ stands for either the reactor level, the reactor pressure, the reactor cooling water temperature or the reactor temperature.

8.5 Experimentation

In the following sections we introduce the results of the four models developed for the reactor level, the reactor pressure, the reactor cooling water temperature and the reactor temperature. Two methods are used. They are the model-based least square estimation method and the fuzzy logic technique.

Three hundred measurements were available for each case study. Two hundred measurements were used as the training data set for building the model. Later on, the remaining hundred measurements were used for testing the model. The evaluation/minimization criterion considered to compute the

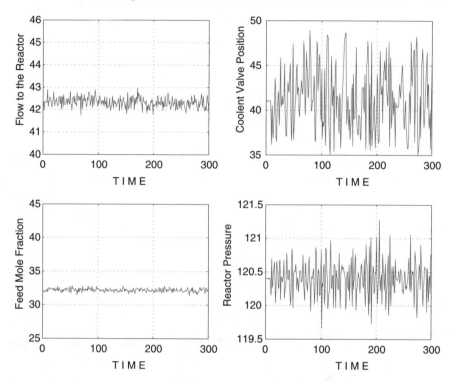

Fig. 8.2. The Input and Output Data for The Reactor Pressure

error difference between the observed output and the developed (LSE/Fuzzy) model outputs selected as the Variance-Accounted-For (VAF).

The Variance-Accounted-For (VAF) was computed over the full three hundred measurements, the training and testing data set.

$$VAF = 1 - \frac{var(y - \hat{y})}{var(y)} \times 100\% \qquad (8.18)$$

Using LSE estimation we built two simple linear model structures. The first model, Model 1, has five parameters and the second one, Model 2, has fifteen parameters based the Volterra time series [33]. The fifteen parameters considered all possible combination of the four input variables. LSE used to compute the model parameters in each sub-problem. The values of the parameters and the structure of the models are reported.

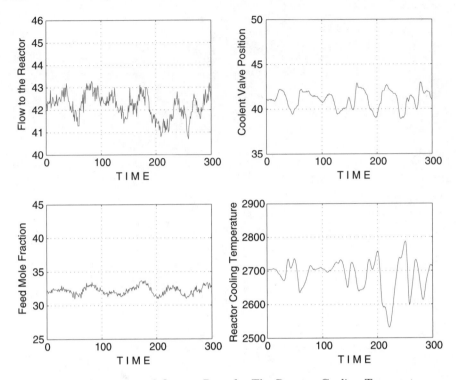

Fig. 8.3. The Input and Output Data for The Reactor Cooling Temperature

We used the Fuzzy Model Identification (FMID) Toolbox implemented in MATLAB [18, 34] to build fuzzy models for each sub-problem. The core of the toolbox is the Gustanfson-Kessel (GK) algorithm.

8.6 Least Squares Modeling Method

In the following four sections, the results for the sub-problems of the reactor are presented. We developed a simple linear model structure with five parameters. The model equation, Model 1, is given as follows:

$$y(k) = a_0 + a_1 u_1(k) + a_2 u_2(k) + a_3 u_3(k) + a_4 y(k-1) \tag{8.19}$$

To achieve a better results in modeling the dynamics of the nonlinear system we have increased the model complexity. A very famous model for nonlinear systems is the Volterra time series.

The Volterra series is a series expansion that consists of linear, bi-linear and tri-linear terms [33]. A Volterra series with chosen order and time delays

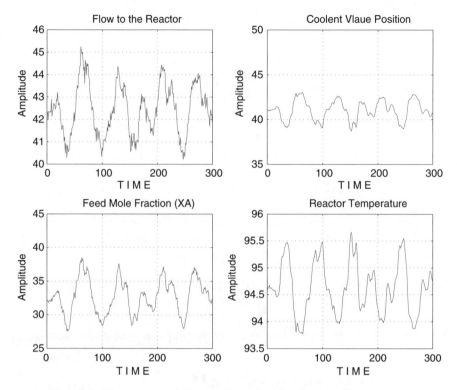

Fig. 8.4. The Input and Output Data for The Reactor Temperature

can be used to model dynamical nonlinear systems. The general description of the Volterra series is given as follows:

$$y(k) = \sum_{i=0}^{\infty} a_i u(k - \tau_i)$$

$$+ \sum_{i=0}^{\infty} \sum_{j=0}^{\infty} b_{ij} u(k - \tau_i) u(k - \tau_j)$$

$$+ \sum_{i=0}^{\infty} \sum_{j=0}^{\infty} \sum_{l=0}^{\infty} c_{ijl} u(k - \tau_i) u(k - \tau_j) u(k - \tau_l)$$

$$+ \cdots \tag{8.20}$$

The data vector u represents the input signal, the vector y represents the model output, the vector a represents the coefficient vector and τ_i, τ_j, τ_l represent the different system delay elements.

The infinite Volterra series can be truncated to lower order for a stable systems. Thus, (8.20) can be simplified as follows:

$$y(k) = \sum_{i=0}^{N-1} a_i u(k - \tau_i)$$

$$+ \sum_{i=0}^{N-1} \sum_{j=0}^{N-1} b_{ij} u(k - \tau_i) u(k - \tau_j)$$

$$+ \sum_{i=0}^{N-1} \sum_{j=0}^{N-1} \sum_{l=0}^{N-1} c_{ijl} u(k - \tau_i) u(k - \tau_j) u(k - \tau_l)$$

$$+ \cdots \tag{8.21}$$

N is the number of collected measurements. Most of the studies were limited to the second or third order Volterra series representation [33, 35, 36].

In our case study, we adopted the second order Volterra time-series to model the reactor subsystems. The model equation, Model 2, is given as follows:

$$\begin{aligned}
y(k) = {} & a_0 + a_1 u_1(k) + a_2 u_2(k) + a_3 u_3(k) + a_4 y(k-1) \\
& + b_1 u_1^2(k) + b_2 u_2^2(k) + b_3 u_3^2(k) + b_4 y^2(k-1) \\
& + b_5 u_1(k) u_2(k) + b_6 u_1(k) u_3(k) + b_7 u_1(k) y(k-1) \\
& + b_8 u_2(k) u_3(k) + b_9 u_2(k) y(k-1) + b_{10} u_3(k) y(k-1) \quad (8.22)
\end{aligned}$$

8.6.1 Modeling Reactor Level Using LSE

Using LSE, the regression model parameters for the reactor level were estimated for both Model 1 and Model 2. The characteristics of the output from Model 2 is shown in Fig. 8.5. A comparison between the computed VAF for both cases is given in Table 8.1.

From Table 8.1 we can see that increasing the number of model parameters helped in deducing the error difference between the observed and estimated model responses.

The equation which describes Model 2 for the reactor level using LSE is given as follows:

$$\begin{aligned}
y(k) = {} & -406.2761 + 12.8524 u_1(k) + 1.2111 u_2(k) - 1.7364 u_3(k) \\
& + 5.2080 y(k-1) + 0.204 u_1^2(k) + 0.1091 u_2^2(k) + 0.0848 u_3^2(k) \\
& + 0.0062 y^2(k-1) - 0.1861 u_1(k) u_2(k) - 0.1908 u_1(k) u_3(k) \\
& - 0.2145 u_1(k) y(k-1) - 0.0837 u_2(k) u_3(k) \\
& + 0.0046 u_2(k) y(k-1) + 0.1 u_3(k) y(k-1) \tag{8.23}
\end{aligned}$$

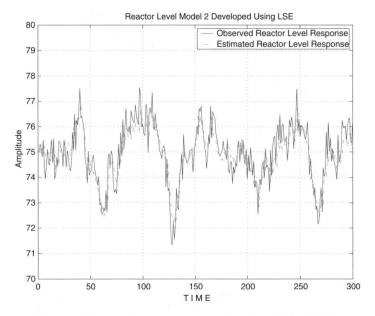

Fig. 8.5. Observed Reactor Level and Regression Model (Model 2) Responses

Table 8.1. Reactor Level Computed VAF

	Model 1	Model 2
VAF	53.4567	68.7256

8.6.2 Modeling Reactor Pressure Using LSE

Figure 8.6 show the model developed for the reactor pressure based LSE. The values of the VAF for both Model 1 and Model 2 are given in Table 8.2. From Table 8.2 it is shown that Model 2 produced an output response with a better VAF.

$$y(k) = 2347 - 66u_1(k) + 2.1u_2(k) + 5.7u_3(k) - 14.3y(k-1)$$
$$+ 0.3u_1^2(k) - 0.1u_3^2(k) + 0.3u_1(k)y(k-1) \tag{8.24}$$

Many of the fifteen model parameters found to have a zero coefficient for many of its variables. The model reduced to a very simple structure as given in (8.24).

8.6.3 Modeling Reactor Cooling Temperature Using LSE

An input-output fifteen parameter model for the reactor cooling temperature was developed. The model parameters were estimated using LSE. From

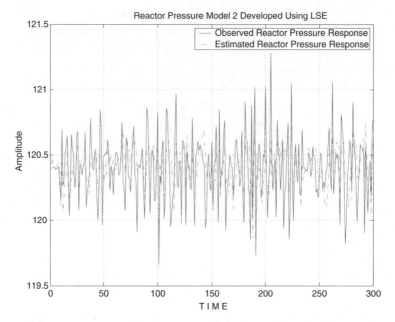

Fig. 8.6. Observed Reactor Pressure and Regression Model (Model 2) Responses

Table 8.2. Reactor Pressure Computed VAF

	Model 1	Model 2
VAF	31.5847	41.1551

the computed results, it was found that the effect of feedback on the model structure, Model 1, can be neglected. Since the coefficient multiplied by the feedback term $y(k-1)$ was almost zero. Model 1 is given in (8.25).

$$y(k) = 2559.5 + 33.6u_1(k) - 25.7u_2(k) - 7.3u_3(k) \qquad (8.25)$$

The developed model for the reactor cooling water temperature, Model 2, is given in (8.26).

$$\begin{aligned}
y(k) = {}& 5728 - 265.6u_1(k) + 118.7u_2(k) - 37.6u_3(k) - 0.5y(k-1) \\
& + 3.1u_1^2(k) - 2.1u_2^2(k) + 1.8u_3^2(k) + 5.5u_1(k)u_2(k) \\
& - 8.6u_1(k)u_3(k) + 2.1u_2(k)u_3(k) - 0.1u_2(k)y(k-1) \\
& + 0.1u_3(k)y(k-1)
\end{aligned} \qquad (8.26)$$

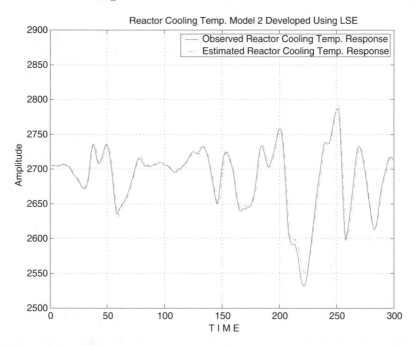

Fig. 8.7. Observed Reactor Cooling Temp. and Regression Model (Model 2) Responses

Table 8.3. Reactor Cooling Temperature Computed VAF

	Model 1	Model 2
VAF	71.0136	96.7729

The observed and estimated outputs for Model 2 of the reactor pressure is shown in Fig. 8.7. A comparison between the computed VAF for both cases is given in Table 8.3. It can be seen from the Table that there is a significant improvement in the VAF for Model 2.

8.6.4 Modeling Reactor Temperature Using LSE

It is important to remind that the models we are building for various sub-problems need to be simple in structure. There is no need to increase model complexity as long as a simple model structure with few number of parameters can satisfy our objective of having good modeling capabilities.

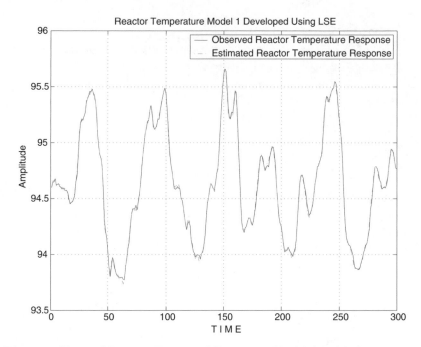

Fig. 8.8. Observed Reactor Temp. and Regression Model (Model 1) Responses

The computed VAF for Model 1 and Model 2 were 99.8344 and 99.9294, respectively. From the developed results, we found that Model 1 output had a high VAF. This is why we provide the equation of Model 1 since it is more simpler. This model is given in (8.27). The characteristics of the developed model, Model 1, is shown in Fig. 8.8.

$$y(k) = 112.3544 - 0.0021u_1(k) - 0.4251u_2(k)$$
$$- 0.0061u_3(k) + 0.0001y(k-1) \qquad (8.27)$$

8.7 Fuzzy Logic Modeling Method

Using the FMID Matlab Toolbox [17, 18] we developed fuzzy models for the reactor level, reactor pressure, reactor cooling temperature and reactor temperature. To achieve a higher VAF, cluster numbers σ, is proposed to take values between 1 to 10. The values of the VAF were computed in each cluster number case and reported in tables. A fuzzy model of the Tennessee Eastman chemical reactor subsystems can be described using a set of rules as given in (8.28).

$$R_i : \textbf{If } y(k-1) \textbf{ is } A_1 \textbf{ and } u_1(k) \textbf{ is } B_1 \textbf{ and } u_2(k) \textbf{ is } B_2$$
$$\textbf{and } u_3(k) \textbf{ is } B_3 \textbf{ then } y(k) \textbf{ is } c_i \tag{8.28}$$

8.7.1 Modeling Reactor Level Using FL

For the process under study, we have three main inputs $u_1(k), u_2(k), u_3(k)$ stand for the flow to the reactor, coolant valve position and the feed mole fraction. The measured output $y(k)$ in this case is the reactor level. Figure 8.9 show the observed reactor response and the response using the fuzzy model. The membership function characteristics for the developed model are given in Fig. 8.10. Table 8.4 shows the cluster values and the corresponding VAF values. The best VAF computed at $\sigma = 8$. The set of generated rules using fuzzy logic computed are given. In Table 8.5, we show the values of the cluster centers.

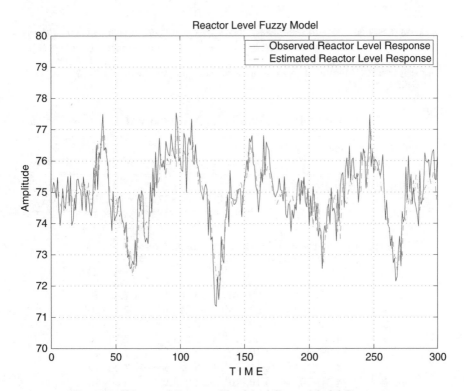

Fig. 8.9. Observed Reactor Level and Fuzzy Model Responses

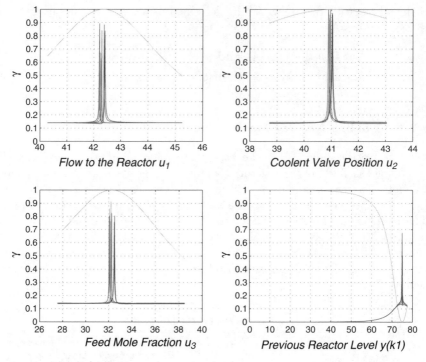

Fig. 8.10. Membership Functions for the Reactor Level Fuzzy Model

Table 8.4. Reactor Level Computed VAF

Number of Clusters	1	2	3	4	5
VAF	53.4567	67.2716	67.1167	67.5607	67.4991

Number of Clusters	6	7	8	9	10
VAF	68.2710	66.1142	**68.5872**	66.6305	62.5936

Table 8.5. Reactor Level Fuzzy Model: Cluster Centers

Rule	u_1	u_2	u_3	$y(k-1)$
1	$4.22 \cdot 10^1$	$4.09 \cdot 10^1$	$3.20 \cdot 10^1$	$7.51 \cdot 10^1$
2	$4.22 \cdot 10^1$	$4.09 \cdot 10^1$	$3.21 \cdot 10^1$	$7.50 \cdot 10^1$
3	$4.23 \cdot 10^1$	$4.09 \cdot 10^1$	$3.22 \cdot 10^1$	$7.50 \cdot 10^1$
4	$4.23 \cdot 10^1$	$4.09 \cdot 10^1$	$3.22 \cdot 10^1$	$7.50 \cdot 10^1$
5	$4.23 \cdot 10^1$	$4.10 \cdot 10^1$	$3.23 \cdot 10^1$	$7.50 \cdot 10^1$
6	$4.24 \cdot 10^1$	$4.11 \cdot 10^1$	$3.22 \cdot 10^1$	$7.22 \cdot 10^0$
7	$4.24 \cdot 10^1$	$4.10 \cdot 10^1$	$3.25 \cdot 10^1$	$7.49 \cdot 10^1$
8	$4.24 \cdot 10^1$	$4.10 \cdot 10^1$	$3.25 \cdot 10^1$	$7.49 \cdot 10^1$

1. **If u_1 is A_{11} and u_2 is A_{12} and u_3 is A_{13} and $y(k-1)$ is A_{14} then**
$y(k) = -3.29 \cdot 10^2 u_1 - 1.04 \cdot 10^2 u_2 - 1.74 \cdot 10^2 u_3 - 2.69 \cdot 10^1 y(k-1) +$
$2.60 \cdot 10^4$

2. **If u_1 is A_{21} and u_2 is A_{22} and u_3 is A_{23} and $y(k-1)$ is A_{24} then**
$y(k) = 5.65 \cdot 10^2 u_1 - 5.89 \cdot 10^1 u_2 + 1.01 \cdot 10^2 u_3 - 2.36 \cdot 10^2 y(k-1) -$
$7.08 \cdot 10^3$

3. **If u_1 is A_{31} and u_2 is A_{32} and u_3 is A_{33} and $y(k-1)$ is A_{34} then**
$y(k) = 7.93 \cdot 10^2 u_1 + 5.56 \cdot 10^2 u_2 - 2.52 \cdot 10^2 u_3 + 4.10 \cdot 10^2 y(k-1) -$
$7.90 \cdot 10^4$

4. **If u_1 is A_{41} and u_2 is A_{42} and u_3 is A_{43} and $y(k-1)$ is A_{44} then**
$y(k) = -1.28 \cdot 10^3 u_1 - 2.40 \cdot 10^2 u_2 + 3.32 \cdot 10^2 u_3 + 2.77 \cdot 10^2 y(k-1) +$
$3.29 \cdot 10^4$

5. **If u_1 is A_{51} and u_2 is A_{52} and u_3 is A_{53} and $y(k-1)$ is A_{54} then**
$y(k) = 1.70 \cdot 10^2 u_1 - 9.01 \cdot 10^1 u_2 + 6.19 \cdot 10^1 u_3 - 4.80 \cdot 10^2 y(k-1) +$
$3.06 \cdot 10^4$

6. **If u_1 is A_{61} and u_2 is A_{62} and u_3 is A_{63} and $y(k-1)$ is A_{64} then**
$y(k) = 3.69 \cdot 10^1 u_1 + 1.66 \cdot 10^1 u_2 - 2.84 \cdot 10^1 u_3 + 2.76 \cdot 10^{-1} y(k-1) -$
$1.25 \cdot 10^3$

7. **If u_1 is A_{71} and u_2 is A_{72} and u_3 is A_{73} and $y(k-1)$ is A_{74} then**
$y(k) = -3.56 \cdot 10^2 u_1 - 4.71 \cdot 10^2 u_2 + 1.52 \cdot 10^2 u_3 + 2.43 \cdot 10^2 y(k-1) +$
$1.14 \cdot 10^4$

8. **If u_1 is A_{81} and u_2 is A_{82} and u_3 is A_{83} and $y(k-1)$ is A_{84} then**
$y(k) = 4.39 \cdot 10^2 u_1 + 4.04 \cdot 10^2 u_2 - 2.20 \cdot 10^2 u_3 - 1.83 \cdot 10^2 y(k-1) -$
$1.43 \cdot 10^4$

8.7.2 Modeling Reactor Pressure Using FL

The fuzzy model for the reactor pressure was developed using Gustanfson-Kessel (GK) algorithm based the FMID Matlab Toolbox. Figure 8.11 and Fig. 8.12 show the observed/estimated reactor pressure responses and the membership function characteristics for the developed model. Table 8.6 shows the values of the cluster centers for the developed fuzzy model. The set of generated rules are given with $\sigma = 10$.

1. **If u_1 is A_{11} and u_2 is A_{12} and u_3 is A_{13} and $y(k-1)$ is A_{14} then**
$y(k) = 9.82 \cdot 10^1 u_1 - 3.90 \cdot 10^0 u_2 - 1.66 \cdot 10^2 u_3 - 9.77 \cdot 10^{-2} y(k-1) +$
$1.48 \cdot 10^3$

2. **If u_1 is A_{21} and u_2 is A_{22} and u_3 is A_{23} and $y(k-1)$ is A_{24} then**
$y(k) = 2.57 \cdot 10^2 u_1 + 1.65 \cdot 10^2 u_2 - 1.96 \cdot 10^3 u_3 - 2.16 \cdot 10^3 y(k-1) +$
$3.06 \cdot 10^5$

3. **If u_1 is A_{31} and u_2 is A_{32} and u_3 is A_{33} and $y(k-1)$ is A_{34} then**
$y(k) = -3.92 \cdot 10^1 u_1 - 6.81 \cdot 10^2 u_2 + 6.24 \cdot 10^3 u_3 + 7.24 \cdot 10^3 y(k-1) -$
$1.04 \cdot 10^6$

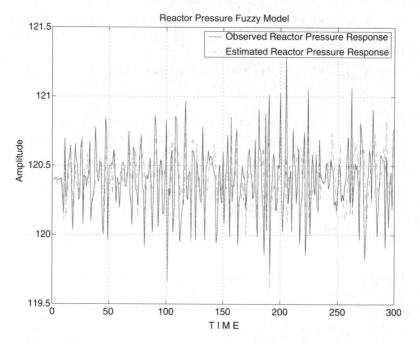

Fig. 8.11. Observed Reactor Pressure and Fuzzy Model Responses

4. **If** u_1 **is** A_{41} **and** u_2 **is** A_{42} **and** u_3 **is** A_{43} **and** $y(k-1)$ **is** A_{44} **then**
   $y(k) = -4.47 \cdot 10^3 u_1 + 2.27 \cdot 10^3 u_2 + 1.18 \cdot 10^3 u_3 - 1.31 \cdot 10^4 y(k-1)+$
   $1.64 \cdot 10^6$

5. **If** u_1 **is** A_{51} **and** u_2 **is** A_{52} **and** u_3 **is** A_{53} **and** $y(k-1)$ **is** A_{54} **then**
   $y(k) = 4.05 \cdot 10^3 u_1 - 2.18 \cdot 10^3 u_2 - 2.62 \cdot 10^3 u_3 + 1.15 \cdot 10^4 y(k-1)-$
   $1.38 \cdot 10^6$

6. **If** u_1 **is** A_{61} **and** u_2 **is** A_{62} **and** u_3 **is** A_{63} **and** $y(k-1)$ **is** A_{64} **then**
   $y(k) = 3.34 \cdot 10^2 u_1 - 1.80 \cdot 10^2 u_2 + 8.49 \cdot 10^2 u_3 + 1.65 \cdot 10^3 y(k-1)-$
   $2.32 \cdot 10^5$

7. **If** u_1 **is** A_{71} **and** u_2 **is** A_{72} **and** u_3 **is** A_{73} **and** $y(k-1)$ **is** A_{74} **then**
   $y(k) = -1.96 \cdot 10^3 u_1 - 1.11 \cdot 10^2 u_2 - 2.00 \cdot 10^3 u_3 - 2.79 \cdot 10^3 y(k-1)+$
   $4.88 \cdot 10^5$

8. **If** u_1 **is** A_{81} **and** u_2 **is** A_{82} **and** u_3 **is** A_{83} **and** $y(k-1)$ **is** A_{84} **then**
   $y(k) = 1.94 \cdot 10^3 u_1 + 8.84 \cdot 10^2 u_2 - 2.36 \cdot 10^3 u_3 - 2.98 \cdot 10^3 y(k-1)+$
   $3.16 \cdot 10^5$

9. **If** u_1 **is** A_{91} **and** u_2 **is** A_{92} **and** u_3 **is** A_{93} **and** $y(k-1)$ **is** A_{94} **then**
   $y(k) = -3.09 \cdot 10^2 u_1 - 4.39 \cdot 10^2 u_2 + 1.99 \cdot 10^3 u_3 + 1.36 \cdot 10^3 y(k-1)-$
   $1.96 \cdot 10^5$

10. **If** u_1 **is** A_{101} **and** u_2 **is** A_{102} **and** u_3 **is** A_{103} **and** $y(k-1)$ **is** A_{104} **then**
    $y(k) = 1.93 \cdot 10^2 u_1 + 2.70 \cdot 10^2 u_2 - 1.30 \cdot 10^3 u_3 - 6.86 \cdot 10^2 y(k-1)+$
    $1.05 \cdot 10^5$

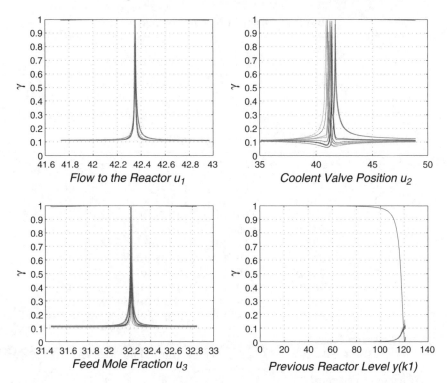

Fig. 8.12. Membership Functions for the Reactor Pressure Fuzzy Model

8.7.3 Modeling Reactor Cooling Temperature Using FL

To develop a fuzzy model for the reactor cooling temperature, we used the FMID Matlab Toolbox with various cluster numbers taking values from $1, \ldots, 10$. The results are shown in Table 8.7, Table 8.8 and Table 8.9. The best

Table 8.6. Reactor Pressure Fuzzy Model: Cluster Centers

Rule	u_1	u_2	u_3	$y(k-1)$
1	$4.23 \cdot 10^1$	$4.11 \cdot 10^1$	$3.22 \cdot 10^1$	$2.14 \cdot 10^0$
2	$4.23 \cdot 10^1$	$4.10 \cdot 10^1$	$3.22 \cdot 10^1$	$1.20 \cdot 10^2$
3	$4.23 \cdot 10^1$	$4.13 \cdot 10^1$	$3.22 \cdot 10^1$	$1.20 \cdot 10^2$
4	$4.23 \cdot 10^1$	$4.11 \cdot 10^1$	$3.22 \cdot 10^1$	$1.20 \cdot 10^2$
5	$4.23 \cdot 10^1$	$4.11 \cdot 10^1$	$3.22 \cdot 10^1$	$1.20 \cdot 10^2$
6	$4.23 \cdot 10^1$	$4.09 \cdot 10^1$	$3.22 \cdot 10^1$	$1.20 \cdot 10^2$
7	$4.24 \cdot 10^1$	$4.14 \cdot 10^1$	$3.22 \cdot 10^1$	$1.20 \cdot 10^2$
8	$4.24 \cdot 10^1$	$4.14 \cdot 10^1$	$3.22 \cdot 10^1$	$1.20 \cdot 10^2$
9	$4.24 \cdot 10^1$	$4.17 \cdot 10^1$	$3.22 \cdot 10^1$	$1.20 \cdot 10^2$
10	$4.24 \cdot 10^1$	$4.17 \cdot 10^1$	$3.22 \cdot 10^1$	$1.20 \cdot 10^2$

Table 8.7. Reactor Pressure Computed VAF

Number of Clusters	1	2	3	4	5
VAF	32.5736	42.0433	43.1150	43.1039	44.5870
Number of Clusters	6	7	8	9	10
VAF	44.8082	42.3875	45.0379	45.9072	**46.4975**

Table 8.8. Reactor Cooling Temperature: Cluster Centers

Rule	u_1	u_2	u_3	$y(k-1)$
1	$4.17 \cdot 10^1$	$4.03 \cdot 10^1$	$3.18 \cdot 10^1$	$2.68 \cdot 10^3$
2	$4.18 \cdot 10^1$	$4.15 \cdot 10^1$	$3.21 \cdot 10^1$	$1.99 \cdot 10^3$
3	$4.19 \cdot 10^1$	$3.95 \cdot 10^1$	$3.17 \cdot 10^1$	$2.72 \cdot 10^3$
4	$4.24 \cdot 10^1$	$4.09 \cdot 10^1$	$3.24 \cdot 10^1$	$2.71 \cdot 10^3$
5	$4.24 \cdot 10^1$	$4.11 \cdot 10^1$	$3.19 \cdot 10^1$	$2.70 \cdot 10^3$
6	$4.24 \cdot 10^1$	$4.19 \cdot 10^1$	$3.24 \cdot 10^1$	$2.69 \cdot 10^3$
7	$4.26 \cdot 10^1$	$4.10 \cdot 10^1$	$3.26 \cdot 10^1$	$2.71 \cdot 10^3$
8	$4.26 \cdot 10^1$	$4.22 \cdot 10^1$	$3.31 \cdot 10^1$	$2.65 \cdot 10^3$
9	$4.27 \cdot 10^1$	$4.09 \cdot 10^1$	$3.22 \cdot 10^1$	$2.71 \cdot 10^3$
10	$4.27 \cdot 10^1$	$4.18 \cdot 10^1$	$3.25 \cdot 10^1$	$2.68 \cdot 10^3$

Table 8.9. Reactor Cooling Temperature Computed VAF

Number of Clusters	1	2	3	4	5
VAF	71.0136	96.7480	94.4091	94.9243	97.1585
Number of Clusters	6	7	8	9	10
VAF	96.3221	97.2173	92.8069	97.6544	**97.7052**

value of the VAF is achieved with numbers of clusters equal to 10. Figure 8.13 shows both the actual output of the reactor cooling temperature model and the estimated output using the fuzzy model. The membership function characteristics for the developed model are shown in Fig. 8.14. Table 8.6 shows the values of the cluster centers for the fuzzy reactor cooling temperature model.

1. **If** u_1 is A_{11} **and** u_2 is A_{12} **and** u_3 is A_{13} **and** $y(k-1)$ is A_{14} **then**
   $y(k) = -9.22 \cdot 10^0 u_1 + 3.15 \cdot 10^0 u_2 + 2.86 \cdot 10^1 u_3 + 1.03 \cdot 10^0 y(k-1) - 7.40 \cdot 10^2$

2. **If** u_1 is A_{21} **and** u_2 is A_{22} **and** u_3 is A_{23} **and** $y(k-1)$ is A_{24} **then**
   $y(k) = -7.34 \cdot 10^5 u_1 - 2.18 \cdot 10^5 u_2 + 4.81 \cdot 10^5 u_3 - 1.13 \cdot 10^2 y(k-1) + 2.46 \cdot 10^7$

3. **If** u_1 is A_{31} **and** u_2 is A_{32} **and** u_3 is A_{33} **and** $y(k-1)$ is A_{34} **then**
   $y(k) = -3.06 \cdot 10^0 u_1 - 1.98 \cdot 10^1 u_2 - 8.50 \cdot 10^0 u_3 + 9.38 \cdot 10^{-1} y(k-1) + 1.36 \cdot 10^3$

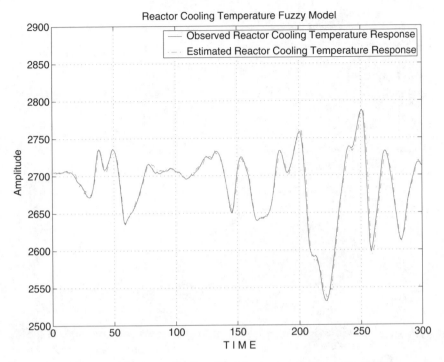

Fig. 8.13. Observed Reactor Cooling Temperature and Fuzzy Model Responses

4. **If u_1 is A_{41} and u_2 is A_{42} and u_3 is A_{43} and $y(k-1)$ is A_{44} then**
   $y(k) = -2.24 \cdot 10^1 u_1 + 1.36 \cdot 10^1 u_2 + 2.48 \cdot 10^1 u_3 + 9.21 \cdot 10^{-1} y(k-1) - 1.96 \cdot 10^2$

5. **If u_1 is A_{51} and u_2 is A_{52} and u_3 is A_{53} and $y(k-1)$ is A_{54} then**
   $y(k) = 3.09 \cdot 10^{-1} u_1 + 2.30 \cdot 10^1 u_2 + 4.79 \cdot 10^0 u_3 + 1.99 \cdot 10^0 y(k-1) - 3.79 \cdot 10^3$

6. **If u_1 is A_{61} and u_2 is A_{62} and u_3 is A_{63} and $y(k-1)$ is A_{64} then**
   $y(k) = 3.28 \cdot 10^1 u_1 - 1.66 \cdot 10^1 u_2 - 1.22 \cdot 10^1 u_3 + 4.75 \cdot 10^{-1} y(k-1) + 1.11 \cdot 10^3$

7. **If u_1 is A_{71} and u_2 is A_{72} and u_3 is A_{73} and $y(k-1)$ is A_{74} then**
   $y(k) = 6.74 \cdot 10^0 u_1 + 9.50 \cdot 10^0 u_2 - 9.71 \cdot 10^0 u_3 + 3.68 \cdot 10^{-1} y(k-1) + 1.35 \cdot 10^3$

8. **If u_1 is A_{81} and u_2 is A_{82} and u_3 is A_{83} and $y(k-1)$ is A_{84} then**
   $y(k) = -1.92 \cdot 10^0 u_1 + 1.13 \cdot 10^1 u_2 - 3.37 \cdot 10^0 u_3 + 1.90 \cdot 10^0 y(k-1) - 2.66 \cdot 10^3$

9. **If u_1 is A_{91} and u_2 is A_{92} and u_3 is A_{93} and $y(k-1)$ is A_{94} then**
   $y(k) = 5.50 \cdot 10^0 u_1 - 8.79 \cdot 10^0 u_2 - 8.90 \cdot 10^0 u_3 + 1.08 \cdot 10^0 y(k-1) + 1.80 \cdot 10^2$

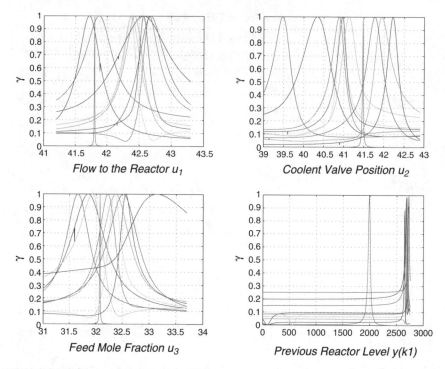

Fig. 8.14. Membership Functions for the Reactor Cooling Temperature Fuzzy Model

10. **If** u_1 is A_{101} **and** u_2 is A_{102} **and** u_3 is A_{103} **and** $y(k-1)$ is A_{104} **then**
$y(k) = 8.94 \cdot 10^0 u_1 - 4.15 \cdot 10^1 u_2 - 1.44 \cdot 10^0 u_3 + 8.42 \cdot 10^{-1} y(k-1) +$
$1.83 \cdot 10^3$

8.7.4 Modeling Reactor Temperature Using FL

From the results developed using the LSE models, for the reactor temperature, it was shown that the five parameter model developed using LSE method was performing well with respect to the VAF criterion. The model had a high VAF. This is why we were expecting that the fuzzy model can be developed using a small number of clusters.

Table 8.10 shows the various number of clusters with the corresponding values of the VAF. Table 8.6 show the values of the cluster centers for the fuzzy reactor temperature model. The four rules which describes the fuzzy model for the reactor temperature are given.

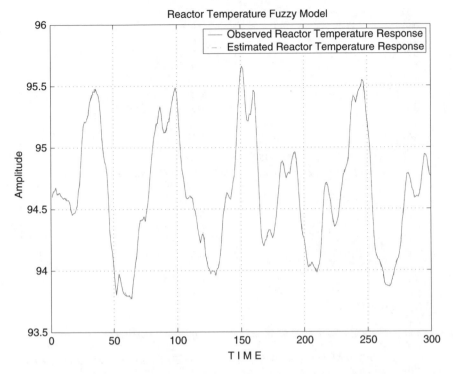

Fig. 8.15. Observed Reactor Temperature and Fuzzy Model Responses

1. **If** u_1 is A_{11} **and** u_2 is A_{12} **and** u_3 is A_{13} **and** $y(k-1)$ is A_{14} **then**
   $y(k) = -9.43 \cdot 10^{-2}u_1 - 9.18 \cdot 10^{-1}u_2 + 3.53 \cdot 10^{-1}u_3 + 7.23 \cdot 10^{-3}$
   $y(k-1) + 1.25 \cdot 10^2$

2. **If** u_1 is A_{21} **and** u_2 is A_{22} **and** u_3 is A_{23} **and** $y(k-1)$ is A_{24} **then**
   $y(k) = 4.36 \cdot 10^{-2}u_1 - 3.66 \cdot 10^{-1}u_2 - 2.14 \cdot 10^{-3}u_3 + 2.71 \cdot 10^{-1}$
   $y(k-1) + 8.23 \cdot 10^1$

Table 8.10. Reactor Temperature Computed VAF

Number of Clusters	1	2	3	4	5
VAF	99.8344	99.9255	99.9303	**99.9341**	99.9315
Number of Clusters	6	7	8	9	10
VAF	99.9313	99.9326	99.9265	99.9330	99.9282

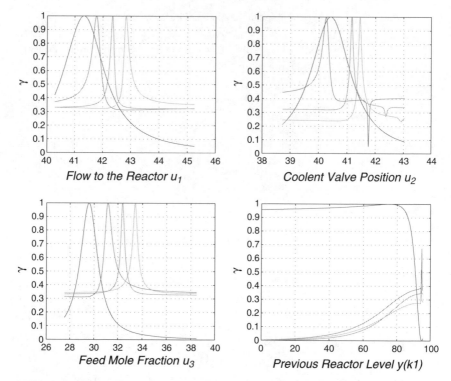

Fig. 8.16. Membership Functions for the Reactor Temperature Fuzzy Model

3. **If** u_1 is A_{31} **and** u_2 is A_{32} **and** u_3 is A_{33} **and** $y(k-1)$ is A_{34} **then**
   $y(k) = -8.72 \cdot 10^{-3} u_1 - 3.66 \cdot 10^{-1} u_2 - 5.37 \cdot 10^{-3} u_3 + 1.37 \cdot 10^{-1}$
   $y(k-1) + 9.73 \cdot 10^{1}$

4. **If** u_1 is A_{41} **and** u_2 is A_{42} **and** u_3 is A_{43} **and** $y(k-1)$ is A_{44} **then**
   $y(k) = -2.77 \cdot 10^{-2} u_1 - 3.50 \cdot 10^{-1} u_2 + 1.07 \cdot 10^{-2} u_3 + 1.58 \cdot 10^{-1}$
   $y(k-1) + 9.49 \cdot 10^{1}$

Table 8.11. Reactor Cooling Temperature: Cluster Centers

Rule	u_1	u_2	u_3	$y(k-1)$
1	$4.13 \cdot 10^{1}$	$4.04 \cdot 10^{1}$	$2.96 \cdot 10^{1}$	$7.56 \cdot 10^{1}$
2	$4.18 \cdot 10^{1}$	$4.03 \cdot 10^{1}$	$3.12 \cdot 10^{1}$	$9.50 \cdot 10^{1}$
3	$4.24 \cdot 10^{1}$	$4.12 \cdot 10^{1}$	$3.24 \cdot 10^{1}$	$9.46 \cdot 10^{1}$
4	$4.28 \cdot 10^{1}$	$4.14 \cdot 10^{1}$	$3.34 \cdot 10^{1}$	$9.45 \cdot 10^{1}$

8.8 Analysis of Results

In this section we compare the results of the models developed using LSE and the developed fuzzy model based Takagi-Sugeno (TS) fuzzy models [17, 18]. Table 8.12 show the computed values of the VAF in all cases considered in this study. It can be seen that the VAF for the fuzzy models outperform the LSE models, in the sub-problems of the reactor pressure, the reactor cooling temperature the reactor temperature. This comparison helps in understanding the nature of the modeling problem of the Tennessee Eastman chemical reactor. This is why it is recommended to use fuzzy logic method as an alternative technique for modeling the dynamics of nonlinear industrial processes.

Table 8.12. The Computed VAF for Model 1, Model 2 and Fuzzy Model

	LSE: Model 1 VAF	LSE: Model 2 VAF	Fuzzy Logic Model VAF
Reactor Level	53.4567	**68.7256**	68.5872
Reactor Pressure	31.5847	41.1551	**46.4975**
Reactor Cooling Temperature	71.0136	96.7729	**97.7052**
Reactor Temperature	99.8344	99.9294	**99.9341**

8.9 Conclusion

In this research, two modeling techniques for building structures of the Tennessee Eastman chemical reactor process were used. They are: (1) the least square estimation; (2) the fuzzy logic based Takagi-Sugeno (TS) algorithm methods. The models developed using the fuzzy logic method were better than the LSE models in the cases of the reactor pressure, the reactor cooling temperature and the reactor temperature. The fuzzy models developed had a better VAF in the above three cases. Compared to other black-box modeling methods, fuzzy modeling has the advantage of providing the mathematical relationship among model variables.

8.10 Acknowledgment

This research was supported by the US-Egypt Science and Technology Joint Fund Program under Grant No. INF4-001-019. Author would like to thank Robert Babuška, Ayman E-Dessouki, Saad Eid, Ahmed Effat and Ghada Khandil for their feedback and comments. A special thanks to Robert Babuška for providing the FMID Matlab Toolbox.

References

1. O. Nelles, *Nonlinear System Identification: From Classical Approaches to Neural Networks and Fuzzy Models*. Springer-Verlag, 2001.
2. J. P. V. Johan A. K. Suykens and B. L. D. Moor, *Artificial Neural Networks for Modelling and Control of Non-Linear Systems*. Kluwer Academic Publishers, Boston, 1995.
3. N. K. P. Magnus Norgaard, Ole Ravn and L. K. Hansen, *Neural Networks for Modelling and Control of Dynamic Systems*. Springer-Verlag, London, 2000.
4. L. Ljung, *System Identification-Theory for the User*. Prentice Hall, Upper Saddle River, N. J., 2 nd edition, 1999.
5. L. Lennart, *System Identification: Theory for the User*. New Jersey: Prentice Hall, 1987.
6. I. J. Leonaritis and S. A. Billings, "Input-output parametric models for nonlinear systems," *International Journal of Control*, vol. 41, pp. 303–344, 1985.
7. R. Haber and L. Keviczky, *Nonlinear System Identification Input-Output Modeling Approach*. Kluwer Academic Publishers, Dordrecht, 1999.
8. T. Wigren, *Recursive Identification Based on the Nonlinear Wiener Model*. Acta Universitatis Upsaliensis: Uppsala, Sweden, 1990.
9. M. Boutayeb and M. Darouach, "Recursive identification method for MISO wiener-hammerstien model," *IEEE Transaction on Neural Networks*, vol. 5, no. 1, pp. 3–14, 1994.
10. D. A. White and D. A. Sofge, *Handbook of Intelligent Control, Neural, Fuzzy, and Adaptive Approaches*. Van Nostrand Reinhold, New York, 1992.
11. R. R. Yager and D. P. Filev, *Essentials of Fuzzy Modeling and Control*. John Wiley, New York, 1994.
12. J. M. Sosa, R. Babuška, and H. B. Verbruggen, "Fuzzy predictive control applied to an air-conditioning system," *Control Engineering Practice*, vol. 5, pp. 1395–1406, 1997.
13. S. Y. Yi and M. J. Chung, "Identification of fuzzy relational model and its application to control," *Fuzzy Sets and Systems*, vol. 59, pp. 25–33, 1993.
14. M. Sugeno and K. Tanaka, "Successive identification of a fuzzy model and its application to prediction of a complex system," *Fuzzy Sets and Systems*, vol. 42, pp. 315–334, 1991.
15. M. 09-Sugeno and T. Yasukawa, "A fuzzy-logic-based approach to qualitative modeling," *IEEE Transactions on Fuzzy Systems*, vol. 1, pp. 7–31, 1993.
16. L. X. Wang, "Fuzzy systems are universal approximators," in *Proceedings of IEE Int. Conf. on Fuzzy Systems, San Diego, USA*, pp. 1163–1170, 1992.
17. R. Babuška, *Fuzzy Modeling for Control*. Klumer Academic Publishers, Boston, 1998.
18. R. Babuška, J. A. Roubos, and H. B. Verbruggen, "Identification of MIMO systems by input-output TS fuzzy models," in *Proceedings of Fuzzy-IEEE'98, Anchorage, Alaska*, 1998.
19. R. Babuška and H. B. Verbruggen, "Identification of composite linear models via fuzzy clustering," in *Proceedings of European Control Conference, Rome, Italy*, pp. 1207–1212, 1995.
20. H. A. Babuška, R. Braake, A. J. Krijgsman, and H. B. Verbruggen, "Comparison of intelligent control schemes for real-time pressure control," *Control Engineering Practice*, vol. 4, pp. 1585–1592, 1996.

21. L. Davis, *Handbook of Genetic Algorithms*. New York: Van Nostrand Reinhold, 1991.
22. M. Phan, J. Juang, and D. Hyland, "On neural networks in identification and control of dynamical systems," tech. rep., 1993.
23. L. Chambers, *Practical Handbook of Genetic Algorithms*. CRC Press, 1995.
24. J. R. Koza, *Genetic Programming III*. MIT Press, 1999.
25. A. Hussian, A. Sheta, M. Kamel, M. Telbany, and A. Abdelwahad, "Modeling of a winding machine using genetic programming," in *Proceedings of the Congress on Evolutionary Computation (CEC2000)*, pp. 398–402, 2000.
26. D. E. Gustafson and W. C. Kessel, "Fuzzy clustering with a fuzzy covariance matrix," in *Proceedings of the IEEE CDC, San Diego, CA, USA*, pp. 761–766, 1979.
27. T. Takagi and M. Sugeno, "Fuzzy identification of systems and its applications to modeling and control," *IEEE Trans. on Systems, Man and Cybernetics*, vol. 15, pp. 116–132, 1985.
28. J. J. Downs and E. F. Vogel, "A plant-wide industrial process control problem," *Computers Chemical Engineering*, vol. 17, pp. 245–255, 1993.
29. N. L. Ricker, "Nonlinear model predictive control of the tennessee eastman challenge process," *Computers Chemical Engineering*, vol. 19, no. 9, pp. 961–981, 1995.
30. N. L. Ricker, "Nonlinear modeling and state estimation of the tennessee eastman challenge process," *Computers Chemical Engineering*, vol. 19, no. 9, pp. 983–1005, 1995.
31. L. Ricker, *http://depts.washington.edu/control/LARRY/TE/*. 1995.
32. N. L. Ricker, "Optimal steady state operation of the tennessee eastman challenge process," *Computers Chemical Engineering*, vol. 19, no. 9, pp. 949–959, 1995.
33. M. Tummala, "Efficient iterative methods for FIR least squares identification," *IEEE Transaction Acoust., Speech, Signal Processing*, vol. 38, no. 5, pp. 887–890, 1990.
34. R. Babuška, *Fuzzy Modeling and Identification Toolbox*. Delft University of Technology, The Netherland, http://lcewww.et.tudelft.nl/bubuska, 1998.
35. S. Nam, S. Kim, and E. Powers, "On the identification of a third-order Volterra nonlinear system using a frequency domain block RLS algorithm," in *Proceedings of the IEEE Transaction Acoust., Speech, Signal Processing*, pp. 2407–2410, 1990.
36. V. Mathews and J. Lee, "A fast recursive least square second-order Volterra filter," in *Proceedings of the IEEE Transaction Acoust., Speech, Signal Processing*, pp. 1383–1386, New York, 1988.

Index

Reviewer List

Printing: Krips bv, Meppel
Binding: Stürtz, Würzburg